T0306091

# MAGMATISM IN THE MCMURDO DRY VALLEYS, ANTARCTICA

The mechanisms of magma movement, chemical differentiation, and physical development are derived from the geochemistry of igneous rocks and from studying exposures of deep magmatic systems that have since solidified and been uplifted and exposed at the Earth's surface. The Ferrar Magmatic System of the McMurdo Dry Valleys in Antarctica provides an unparalleled example of a complete magmatic–volcanic system exposed in unprecedented detail. This book provides an unusual three-dimensional detailed examination of this system, providing insight into many magmatic processes normally unobservable, in particular how basaltic magma moves upwards through the crust, how it entrains, carries, and deposits loads of crystals from great depths, and how this all contributes to Earth's evolution.

Providing an explanation of how magmatic systems operate and how igneous rocks form, this is an invaluable resource ideal for researchers and graduate students in magma physics, igneous petrology, volcanology, and geochemistry.

BRUCE MARSH is Professor Emeritus at Johns Hopkins University. He has spent his career studying the physics and chemistry of planetary magmatic processes, including geothermal energy. He is an elected fellow of the American Geophysical Union, Geological Society of America, Mineralogical Society of America, and the Royal Astronomical Society, and has received some awards.

# MAGMATISM IN THE MCMURDO DRY VALLEYS, ANTARCTICA

BRUCE MARSH

*Johns Hopkins University*

CAMBRIDGE
UNIVERSITY PRESS

Shaftesbury Road, Cambridge CB2 8EA, United Kingdom

One Liberty Plaza, 20th Floor, New York, NY 10006, USA

477 Williamstown Road, Port Melbourne, VIC 3207, Australia

314–321, 3rd Floor, Plot 3, Splendor Forum, Jasola District Centre, New Delhi – 110025, India

103 Penang Road, #05–06/07, Visioncrest Commercial, Singapore 238467

Cambridge University Press is part of Cambridge University Press & Assessment,
a department of the University of Cambridge.

We share the University's mission to contribute to society through the pursuit of
education, learning and research at the highest international levels of excellence.

www.cambridge.org
Information on this title: www.cambridge.org/9781009177085

DOI: 10.1017/9781009177078

First published 2023

*A catalogue record for this publication is available from the British Library*

*Library of Congress Cataloging-in-Publication data*
Names: Marsh, Bruce D., author.
Title: Magmatism in the McMurdo Dry Valleys, Antarctica / Bruce Marsh, Johns Hopkins University.
Description: Cambridge, United Kingdom ; New York, NY : Cambridge University Press, 2022. |
Includes bibliographical references and index.
Identifiers: LCCN 2022030687 (print) | LCCN 2022030688 (ebook) | ISBN 9781009177085 (hardback) |
ISBN 9781009177078 (epub)
Subjects: LCSH: Magmatism–Antarctica–McMurdo Dry Valleys.
Classification: LCC QE461 .M367 2022 (print) | LCC QE461 (ebook) | DDC 551.1/309989–dc23/eng20220919
LC record available at https://lccn.loc.gov/2022030687
LC ebook record available at https://lccn.loc.gov/2022030688

ISBN    978-1-009-17708-5    Hardback

... it will be seen that wherever a dark rock was encountered it proved to be dolerite ...
*(H. T. Ferrar, 1907)*

This is dedicated to the steadfast support over 25 years, both in and out of nine field seasons, by the US National Science Foundation, principally headed by the insightful hand of Scott Borg. And also to the many untold geologists, mainly New Zealanders, whose efforts set the stage for this work.

# Contents

# Figures

# Preface

Magma is the mysterious material thought to play a fundamental role in planetary evolution, leading on Earth somehow to the origin of the dramatic contrast between the Granitic Continents and Basaltic Ocean Basins, not to mention the Oceans and Atmosphere. Despite the countless occurrences of magma as lava, alive and flowing and then solid, and unearthed solidified plutons, it has proven elusive to any simple behavioral understanding. Most of this uncertainty comes from the inherent challenge of being allowed to examine only bits and pieces of its being, starting with formation at depth, ascension, and high-level emplacement or eruption. Around each process detailed ideas and explanations abound, although there is a general paucity of an overall integration.

Many tools are available to examine and reveal the intimate workings of magma on all spatial and chemical scales. Given the opportunity to examine any significant piece of the machinery of the magmatic process, much can be revealed of the life of magma. This is exactly what is reported here; namely, the detailed examination of a beautifully integrated and stunningly exposed magmatic system in one of Earth's most beautiful terrains: the Mars-like vestiges of the McMurdo Dry Valleys of Antarctica.

This book, then, is about the life and intimate, nearly day to day, existential behavior of basaltic magma; especially how deep seated, intrusive magma relates to plutons and lavas; how it differentiates into more evolved compositions, and how it lives within and modifies the harboring granitic continental crust.

To be sure, how basaltic magmas might transform chameleonlike into a spate of subsidiary compositions, quite likely leading to Granitic Continents and Ocean Basins, was in true essence discovered more than a century ago by realizing that simply adding and subtracting indigenous crystals readily produces pretty much any final composition. This is the holy grail of magmatic speciation: differentiation by fractional crystallization. But how exactly does this physically happen? Ideas abound and mountains of evidence are globally evident; some exhibiting majestically layered exotic bewildering deposits. With this evidence came a plethora of ideas and hypotheses on magma behavior during generation, extraction, ascension, and crustal emplacement or eruption, leading to the very differentiation of Earth itself.

Some of the more persuasive and real-life evidence, deeply impressing me, came from the detailed studies of the Hawaiian Lava Lakes in the 1960s and 1970s. Basaltic magma erupting at Kilauea, in flowing across the landscape of the East Rift, filled a series of washtub-like pit craters with 15 to 125 m of lava-magma. As soon as stable crusts formed, USGS (US Geological Survey) researchers began drilling though the crusts measuring temperature, magma/rock composition, viscosity, thermal properties, vesiculation, and the ongoing growth of the crust itself. Analytical models of the cooling process showed it to be perfectly described by simple heat conduction, with measurements agreeing with calculation to within a single degree. Nothing fancy was evidently needed and, in the magma itself, nothing truly dramatic seemed to be taking place; no wholesale differentiation was found. The magma seemed content to just sit, cooling, crystalizing, and solidifying.

Small patches of almost granitic silicic differentiates appeared in the upper crust in local tears and lenses, but the initial main body of basaltic magma, once stripped by rapid settling of any pre-existing large crystals of olivine, was pretty much dynamically nearly inert. Could these lava lakes be examples of actual magma chambers? Or are they simply too small, where nothing much can be expected to happen. Perhaps there is a critical thickness, chamber shape, or longer cooling time that gives rise to the richly varied, exotic, highly layered monuments upon which the central tenets of magmatic differentiation have been founded.

In a concerted effort to understand how the simplest of sheets of basaltic magma cools and crystalizes, and perhaps differentiates into more silicic products, I was taken with the notion of further exploring the lava lake concept by finding similar sheets of basaltic magma, both thick lavas and sills, and studying them in a postmortem fashion like a solidified lava lake. Beginning with lava lake-like thin sheets, like that of the 2 km wide by 70 m thick sheet at Shonkin Sag in Montana, richly differentiated, the search spread. A catalog was made of every sheet ever found, noting its size, shape, degree of exposure, completeness, chemical composition, and distinguishing characteristics. With an eye towards physical evidence for chemical differentiation, it soon became clear that the definitive dividing line between magmas that fractionated into a series of compositions and those that didn't was whether the intruding magma contained large crystals; phenocrysts, xenocrysts, or primocrysts, it didn't matter which. Big crystals can settle rapidly, escaping being caught in the solidification fronts marching inward from all borders. Here it was, again, the simple process of adding or subtracting crystals: fractional crystallization.

As for clear indications of extreme differentiation to the point of producing granitic compositions, all that could be found were persistent scattered lenses of granophyric material, almost granite, but not quite. Yet the physical process of producing these almost granite granophyres, although not exactly evident, seemed tractable, if fine examples, well exposed in three dimensions could be found. In this direction the singular seminal work of Bernard Gunn surfaced, shown to me by Margaret Mangan (who at the time was working on the broadly similar York Haven sill) on the petrology of the basaltic (dolerite) sills of the McMurdo Dry Valleys. Miles and miles of exceptional exposures of huge sheets of basaltic rock showing strong compositional and textural diversity. Here was the place to examine, on the very bottom of Earth.

So, it was the granophyric lenses that initially took us to the McMurdo Dry Valleys, and the highly integrated magmatic magnificence of this huge system only became clear after living within it for two field seasons. That is, frankly speaking, at first look this complex of dolerite sills is so vast and varied that it presents itself in the larger context as a hodge-podge of haphazard crosscutting relations having no special order or systematic development. Casual conversations with others who had made cursory examinations of this system assured me that, in overall development and operation, this was no simple system; even the Sr isotopes were all over the map, messy. Yet the many intimately exposed unusual features were naggingly intriguing.

With Gunn's papers in hand and being able to stand back and look over the place, valley to valley, walking, sampling, and tracing out the sills, my entire perspective changed. A magmatic system suddenly appeared of singular distinction, beautifully integrated, and with all the geologic evidence readily available to unravel its history of initiation, development, and operation. In spanning upwards over about 4 vertical kilometers, it exposes a deep-seated, beautifully layered gabbroic pluton connected to a stack of massive sills culminating into tephra and lava on the surface. The vital key to unlocking the dynamics of the whole system is the presence of a massive concentration of large distinctive pyroxene crystals, mainly orthopyroxene (aka Opx), that serve as tracers enabling the flow and behavior of the magma to be read throughout the system. Gunn talked about these big crystals, but mainly in the context of how their presence or absence dictated the bulk composition through fractional crystallization, which was the governing research spirit of the times in the 1960s. The stage was set, all that needed to be done was to unravel the puzzle and understand it as well as we could. And that this geologic stage was squarely set in a singularly historic polar regime of exploration, involving attempts to reach the South Geographic Pole and the South Magnetic Pole, only made the effort that much more stimulating.

Oddly enough, quite by accident, I had for much of my life unknowingly prepared for Antarctica. I was raised on the shore and ice fields of Lake Superior, where much of the year is truly polar. Wanting to understand snow and ice and explore nearby icy shores and islands, I found much enjoyment in reading the major tomes on polar exploration from the town school library; large dark blue volumes by Vilhjalmur Stefansson, Robert Peary, Peter Freuchen, Robert Scott, Fridtjof Nansen, John Franklin, Ernest Shackleton, Roald Amundsen, Frederick Cook, Richard Byrd, among other explorers, including Lewis and Clark and Daniel Boone. My father, William Marsh, and maternal grandfather, Gordon Earl Steinhoff, also read many of these treatises and often discussed the subject matters. Grandfather, himself, sometimes lived much of the winter on a remote Lake Superior island, fishing through the ice with others using dog teams, sometimes nearly being carried away on ice floes. A friend of his, who lived nearby, was William Bakewell, who had been with Shackleton on the *Endurance*, and he sometimes told me stories that Bakewell had told him about being with Shackleton. Later in graduate school I met geologist Laurence Gould, who had been with Byrd, and he told of finding Amundsen's 1911 depot many years later near Mt. Betty, all still in perfect condition, including Amundsen's note of 6–7 January 1912, telling of reaching the Pole.

And so it was that, when I first arrived on the Ross Ice Shelf in earliest 1993, along with other dreary newcomers, like me, standing at 3 AM in bright sunlight, who pointed and asked: "What is that? And that? And that?" I answered every question, without thinking, until someone asked: "How many times have you been here?" Strangely enough, it was like coming home.

Field work in Antarctica under the operation of the US National Science Foundation is a first-class experience. Having spent many years working in remote difficult volcanic field areas, like the Aleutian Islands, I was used to planning every detail of each field season. Every tent stake, stove, box of crackers, sewing kit, etc., had to be procured, packaged, and sent ahead. All logistics had to be invented and, in the end, a great deal of effort was expended in getting everything into the field, mostly on foot, and simply surviving with no outside contacts whatsoever. The scientific rewards and the ability to visit truly singular places of beauty always more than made up for the efforts.

Research in the McMurdo Dry Valleys is entirely different; it is truly special: no darkness, no rain, no grizzly bears, and the best field gear, food, and helicopters are always available. But the preparation is still lengthy and detailed, for although the terrain is breathtakingly beautiful, it is also potentially stunningly dangerous. From our base camps at Solitary Rocks and Bull Pass we roamed the entire Dry Valleys, much of it on foot, after being put down as a day trip, and other days with 'close support' by helicopter all day.

And, although the southern part of the Dry Valleys was discovered by R. F. Scott's "Discovery" expedition in 1902, the wonderful large expanse of the northern Dry Valleys (Wright Valley, Bull Pass, McKelvey Valley, etc.) had only been known for about 35 years when we went there. Students sometimes asked: "Will we walk in areas where no one has ever walked before." Yes, every day. And although much of the geology had been well sketched out for many years, there was, almost everywhere, an abundance of fresh new discoveries to be made. For here the rocks are "indecently" exposed, with no bushes, no shopping centers, no poison ivy, it is all right there at a level where every ounce of one's geologic training continually comes into play. The terrain is unlike anything elsewhere on Earth, it is a relict terrain that achieved it basic form tens of millions of years ago and, in spite of being up against one of Earth's largest ice caps, has been carved mainly by wind and water, pretty much existing unscathed for all this time. To a geologist, it is like a humanist discovering an ancient Babylonian library perfectly intact. It is simply a place of true wonder.

Doing research in the Dry Valleys is much like, I imagine, doing field work on the Moon. Once you land and set up camp – and it takes experience to place camp in a convenient and safe place – there is so much to look at and look after that it is easy to waste the precious time you have there. And unlike usual field trips and field work in more sedate regions, this work is much more like a military campaign. Some students never fully realize what is at stake, that this isn't a summer field camp or an egalitarian university seminar; but this is much more serious business. Others immediately grasp the gravity and splendor of the place and instantly form a high functioning team, much like, perhaps, a well-oiled group of tribal warriors, complete with sensory deprivation. I always put our camps in the most critical and safe geologic locations possible, where high quality work could be done on foot regardless of any outside logistical support. I had a thorough plan of things to do every day, with a half day off on Sundays. We always got a lot done, but there is so much to see and think about that, I'm afraid, we only scratched the surface.

This book, then, tells a story of science and exploration in one of the most enchanting places on Earth.

# Acknowledgments

Many individuals have given freely of their time, data, and advice. These include Bernard Gunn, who discussed detailed petrology and field relations, including perhaps carrying a few sticks of dynamite to get better exposures; Peter Barrett, who freely discussed and gave me his unpublished mapping in the Dry Valleys; Warren Hamilton generously shared his thin sections and papers; David Elliot kindly entertained many queries, furnished papers, and shared figures; Thomas Fleming shared manuscripts and the Kirkpatrick chemical data; Sam Mukasa gave me all the whole rock data he could find for the Dufek; Peter Larson, for his insight and use of the oxygen data; Ken Foland, for his generous efforts to measure the Sr and Nd isotopes; Mark Rattenbury, who so graciously gave me access and explanatory materials to the rock collection at GNS, Lower Hut, and Nick Mortimer for making thin sections; Simon Cox kindly kept me aware of the GNS reports and maps; Steve White, for sharing his insights and manuscripts on the Coombs-Allan Hills centers; and George Denton, David Marchant, and Brenda Hall, who made the geomorphic story come alive. George was especially helpful early on in giving advice on how and where to set up safe base camps, and David succinctly pointed out, during rare, shared helicopter trips under dire circumstances, the finer points of the landscape architecture. Peter Wylie, with his long-ago Greenland field partner, Colin Bull, opened a useful, touching dialog, and furnished me with Colin's "Innocents in the Dry Valleys."

This work started as a chapter for a book, but under the steady and kind encouragement of Adam Martin it grew into a book in and of itself.

A month-long lecture tour, planned as 4 lectures, but blossoming into 17 lectures from Dunedin to Auckland, sponsored by the four major New Zealand universities and the Mineralogical Society of America, provided an abundance of truly special opportunities, including field trips, to freely discuss Antarctic geology and exploration.

I offer my profound heartfelt thanks to those field partners of the most sterling nature any field man could ever hope for, who always went the extra mile carrying the extra sample, and freely discussing what was at hand, making my life easier, especially Michael Zieg, Amanda Charrier, Taber Hersum, Ryan Currier, Adam Simon, and Dean Peterson, among others, and this includes the helicopter aces, especially Chris Dean and Barry James, also among others, often under the direction of Jack Hawkins, who repeatedly took us via "hot" landings to inaccessible places and retrieved us with grace and aplomb. Paul Morin, Zbigniew Malolepszy, and Bonnie Souter introduced highly useful imaging techniques both in the field and the laboratory. Riley Flanagan-Brown was masterful in building and operating the Mush Column experiments. Tushar Mittal helped in a wide assortment of ways from numerical modeling to experiment design and deployment and valuable discussion. And Kaustubh Patwardhan was kindly helpful in many ways here and there.

Cindy Weeks diligently, steadfastly, and carefully helped on all aspects of managing this research, getting us into and out of the field, curating my field notes, samples, field maps, and photographs. She went beyond the normal limits of endurance. Keith Kaneda also worked relentlessly to get all the samples and thin sections properly photographed and carried out a long series of organizational efforts assisting this research. To each I am deeply appreciative and grateful. Lengthy discussions over the years on lava lakes with Herb Shaw, Thomas Wright, and Rosalind Helz were highly helpful, as were endless dialogs with Ian Carmichael on the behavior of magma and the importance of careful petrography.

Above all, special singular thanks to my long-time field partner, confidant, and friend, Judith A. Marsh, who has made it easy for me to do science by joyously holding it all together and keeping the home fires burning for over 50 years. And similar heartfelt thanks to Hannah and Will Marsh who, in their own enthusiasms, never tired of helping and asking and learning about Antarctica, past, present, and future.

Thanks also to my brother James for being such a stalwart sledging partner in our youth, sister Kay, who was always up for a snowy trek or fossil hunting, and brother Bill, who showed me and splendidly demonstrated that a life of adventure and learning could be made into a wonderful occupation. And cheers to Reds, Ed, and Beerg for setting standards of gentlemanly honor; you know who you are.

To the many untold people in McMurdo and elsewhere who helped in so many, many ways in making my research program successful, I also extend my most earnest gratitude.

Sincere thanks also go to the Cambridge group for taking a chance on this manuscript (Matt Lloyd), expertly shepherding all the pieces together (Sarah

Lambert), marshalling and marvelously setting up the Figures (Sapphire Duveau), and through kind eyes and ears making the manuscript clean and fun (Liz Steele).

Finally, much appreciation goes to the many scientists of all kinds who rendezvous each year in McMurdo, going and coming, freely discussing their latest discoveries, adventures, and fears; for being in McMurdo is a true Voyage of Discovery.

# 1

# The Nature of Magmatism

## 1.1 Introduction

Direct knowledge of the composition and spatial distribution of Earth's interior material comes largely by magmatic processes spread across the planet. Aside from the bulk composition of the dominant or carrier magma, the inclusion of exotic primocrysts, xenocrysts, xenoliths, and sundry other material carries intimate information on the nature of Earth's interior, including source regions, the plumbing system, and the dynamics of the governing magmatic transport process itself. The ever-present challenge is, once given this evidence, to decipher the complete magmatic process such that the reason for the magmatism and all that it contains can be clearly understood in the true context of the physical and chemical processes that it represents. In the drive to understand the nature of the underlying magmatic source material this is most often attempted purely through geochemistry in an aim to identify distinctive source reservoirs, with much less regard for the actual physical process of the magmatism itself as a primary guide. This is generally also necessitated by the paucity of deeply exposed magmatic systems where the intimate details of the magmatic delivery can be readily deciphered. A singular exception to this paucity is the Ferrar Magmatic System of the McMurdo Dry Valleys that provides an unparalleled example of a complete magmatic intrusive system. From this system truly fundamental insight into hitherto poorly understood magmatic processes can be gained. Hence, a more balanced approach to understanding magmatic source regions is attempted here using everything found in the rocks themselves, especially their spatial distribution.

## 1.2 Plutonism and Volcanism

Magmatic systems of all sizes and compositions commonly show patterns of individual eruptive centers distributed along otherwise linear tectonic elements.

Island arcs are perhaps the most distinctive, where, along distances of thousands of kilometers, individual volcanic centers, spaced at 50–70 km, have spontaneously operated, off and on, for millions of years (e.g., Marsh, 1976, 1979a, 1979b, 2015). Even oceanic "hot spot" volcanic centers, like Hawaii, show a similar pattern, although here only a single major center predominates at any time (Clague & Dalrymple, 1987). Rift systems are also similar. Although characteristically forming long narrow or piecewise continuous clefts, the magmatism itself, wherever it can be accurately discerned, is in centralized centers. The Great Dyke of Zimbabwe shows this behavior with individual magmatic centers distributed at distances of 70–100 km along a 550 km narrow massive trough, reminiscent of a nascent oceanic ridge (e.g., Worst, 1960; Wilson, 1996). Ocean ridges themselves also show this behavior particularly well (e.g., Schouten et al., 1985; MacDonald, 1986, 2019; Sinton & Detrick, 1992), with magmatic centers operating at discrete intervals along markedly segmented ridges, often existing midway between offsets or other major faults, including Transform Faults. Ascending magma pools at these locations, forming high-level Axial Magma chambers as relatively thin (~100 m) sills, from which magma is dispersed along the ridge. At ridge areas starved of magma, the newly forming oceanic crust is thin and the underlying lithosphere is tectonically emplaced to satisfy the requirements of mass conservation governing the overall flow. All aspects of the ridge, from petrological to biological, morphological, and thermal, are influenced by this segmentation (see the collection of papers in MacLeod et al., 1996). The Ferrar magmatic system of Antarctica may also show something of this pattern of magma ascent and dispersal from highly localized centers, but even more important, occasionally, due to the labors of erosion and relict terrain preservation, a rare magmatic center will reveal the detailed workings of the fully integrated magmatic system itself.

Although all volcanic and near-surface magmatic systems are certainly highly integrated, both physically and dynamically, from source to pluton or volcano, it has proven difficult to gain an intimate understanding of this integration because of limited vertical exposure in Earth's crust. And suites of samples from volcanic and plutonic terrains are individually, in and of themselves, of little help. Volcanic suites sometimes yield a wide range of compositional diversity along with important quenched or incomplete textural development, thereby capturing the element of time, but the spatial context, or provenance, from whence each mass of lava came within the magmatic system, is unknown. There is, no doubt, that a record of the nature and dynamics of the system intimately exists in the pile of lavas simply from stratigraphic position, volume, and composition, but there is no clear way to unravel and relate this accurately to the spatial configuration of the underlying plumbing system and, moreover, how this relates to the process producing the magma.

Plutonic suites, on the other hand, do offer excellent spatial control, but there is no record of the magmatic time series of inputs and outputs leading to establishment and evolution of the final system itself. The rock textures themselves to varying degrees have commonly been seriously thermally annealed, thus obscuring or removing information critical to understanding the history of emplacement and subsequent crystallization. The spatial and temporal record of the dynamics of operation of the system as a whole, relating the vat of magma to its source and to coeval volcanism, has been lost. The all-important initial conditions of the magma that gave rise to the end product are no longer extractable. The prime fundamental questions cannot be answered, namely: What was the style and duration of filling? What was the compositional variation and phenocryst content of the individual magmatic pulses? If phenocrysts and primocrysts were present, where did they come from? Were they sorted prior to injection during ascension or after emplacement into a vat or chamber? And, perhaps above all, how are the dynamics of volcanic and plutonic systems connected?

Purely geochemical petrogenetic scenarios manifestly suffer from this limited understanding by mainly commonly employing, of necessity, only two key end-member regions in considering the production and chemical evolution of magma. These are a source or deep-seated region and a high level, near surface, or supposed magma chamber region. Although it is well appreciated that the intervening pathway or mush column connecting these end points is certainly as important as the endpoints themselves in influencing magma evolution, it has proven elusive to develop a detailed and realistic understanding of this region of magma transfer (e.g., Davidson et al., 2005a, 2005b, 2007). Properties of the magma attributable to the integrated system are thus of necessity collapsed into the endpoints of the system, which reinforces the concept of a virtual magmatic system based on assumed "reservoirs." That is, the detailed isotopic and trace element suites tend to identify characteristic "source" reservoirs, such as deep mantle plumes, fertile or depleted mantle, each of which may have suffered intricate histories of metasomatism due to subduction or other processes, that through mixing and/or interaction with a primary magma are able to match the observed chemical fingerprints. But the proposed scenarios of magmatic processes are often highly unrealistic. Although this approach is also partly employed herein, it is greatly subordinate to the actual process of magmatism as recorded in the rocks themselves. Here, an age-old dictum is adhered to: The in-situ rocks are the ultimate and final court of appeal.

## 1.3 Mechanics of Magmatism

To set a framework for all that follows it is convenient to have in mind some specific magmatic systems that may encapsulate some of what has already been

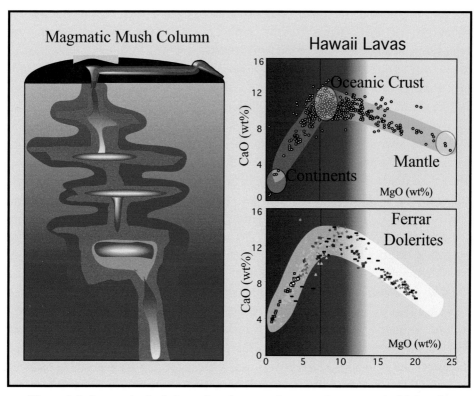

Figure 1.1 A generic depiction of an integrated magmatic system (a Magmatic Mush Column) beneath a volcanic center as a series of high aspect ratio interconnected magma chambers. The color throughout these figures portrays magma temperature, red cooler and yellow hottest; the small black squares amongst a green background depict large crystals of either olivine or orthopyroxene. The spectrum of magma compositions within the system is characterized in a plot of CaO versus MgO (wt.%) for lavas from Hawaii. The first order degree of primitiveness can be gauged by MgO content from typical mantle to oceanic crust to continental material. This variation found within the Ferrar Magmatic System is also shown. The slight difference in slope on the high MgO reflects the role of olivine at Hawaii versus orthopyroxene (Opx) in the Ferrar (after Marsh, 2004).

described. Two magmatic systems serve this purpose: First, a Magmatic Mush Column, which may operate under large centralized magmatic centers like Hawaii (Figure 1.1), and second is an Ocean Ridge Magmatic System, which may exemplify magmatic behavior under active ocean ridges (Figure 1.2).

### 1.3.1 Magmatic Mush Columns

A Magmatic Mush Column (MMC) is an extensive, vertically interconnected stack of sheets and chambers extending upward through the lithosphere and crust,

## Ocean Ridge Magmatic System

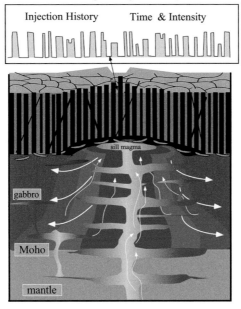

Figure 1.2 The mush column magmatic system at ocean ridges. The system is characterized by a thin sill (Axial Magma Chamber, AMC) capping a vertically extensive system of thin sills and conduits within a massive column of mush. The sheeted dike complex at the top records the history of extraction of magma from the AMC as the plates suddenly pull apart over time periods depicted at the top. Note the small plagiogranite lenses within the upper gabbros formed by solidification front instability (after Marsh, 2015).

capped by a volcanic center (e.g., Marsh, 2004, 2015). Highly fractionated and primitive magmas, and everything in between, may coexist as, respectively, pools of nearly crystal-free magma, crystal-rich magma, thick beds of cumulates, and open or congested (by cognate and wall debris) conduits. Dynamic solidification fronts sheath all system boundaries, advancing inward in response to local geometry, thermal regime, and level of magmatic activity. The eruptive chemical and petrographic nature hinges on the strength and duration of the eruption flux: The stronger and longer the flux, the more crystal-laden and primitive the eruptive. The surficial chemical impression of the deeper nature of the system depends heavily on the temporal dynamics of the system. A system may thus appear petrologically distinct from one episode to the next. The conceptual philosophy behind the measures and processes used to gauge the system strongly colors how the system itself is perceived: "There are many differences and complications in the natural magma in the matter of details, but it is clear that the broad scheme is well understood and that crystallization is the sole control" (Bowen, 1915). That is, Bowenian thinking, for example, yields Bowenian processes.

A key to appreciating the level of physical and chemical integration of a MMC is finding a vertically extensive and intricately exposed system, revealing the spatial connections between lavas, sills, feeders, and ultramafic-layered rocks. It is something of this basic nature that may well be exemplified by the Ferrar dolerites of the McMurdo Dry Valleys, but not entirely. A characterization of the intimate degree of physical and chemical integration of MMC systems is depicted by the basic chemical variation between CaO and MgO of Hawaii and the Ferrar system, as suggested by Figure 1.1. They resemble one another quite closely, even though Hawaiian compositions are controlled by the loss and gain of olivine, whereas the Ferrar is controlled largely by orthopyroxene. The Hawaii compositions all come from lavas whose genetic and spatial origin within the associated MMC is unknown, but, as will become abundantly clear, for the Ferrar the spatial context of each sample is intimately known. Hence the fundamental value of the Ferrar.

This variation between CaO and MgO, in an essential way, depicts the genetic connection between the deeper mantle (MgO ~30 wt.%), the oceanic crust (MgO ~7%), and the highly refined continental crust (MgO ~2%), which is commonly assumed to take place by, at the onset, partial melting followed by fractional crystallization. The fine line at 7% MgO marks the point at the leading edge of the Solidification Front (see Chapter 4) where the phase equilibria changes from single phase saturation, solely olivine or orthopyroxene, to multiple saturation with the appearance of plagioclase and other phases.

### *1.3.2 Ocean Ridge Magmatism*

For much of the past century magma chambers were envisioned to be giant vessels or vats of magma within which crystallization took place throughout. In concert with this classical concept, vast reservoirs of magma were expected at the level of the oceanic crust beneath ocean ridges. When these features were sought using seismic methods, they were not found. Instead, thin (50–100 m), wide (2–3 km) ribbons of magma, or sills, were found (Sinton & Detrick, 1992). This spawned a general model of ridge magmatism more consistent with what is found for systems like the Ferrar, albeit on a smaller scale, and the structure of the oceanic crust is an intimate reflection of the combined process of magmatism in response to plate tectonics.

A general depiction of this system (Figure 1.2) stems from the general form of magmatic sill complexes found in the Ferrar system, in old continental crust, in opiolites (e.g., Nicolas, 1995), and in 3D multi-channel seismic studies in the North Atlantic (e.g., Cartwright & Hansen, 2006), and it is also consistent with what should be expected on a mechanical basis in this tectonic region (e.g., Teagle et al., 2012). This is a magmatic mush column operating under quasi-steady state conditions in response to upwelling and lateral flow (i.e., rifting) due to large-scale mantle convection. The mush column (Figure 1.1) is truncated and stands in partially molten ultramafic mantle rock modally dominated by olivine with subordinate orthopyroxene, clinopyroxene, and plagioclase, with spinel and pyrope-rich garnet at successive higher pressures. At deeper levels (~50–100 km) significant partial melting due to adiabatic decompression (e.g., McKenzie, 1984) forms an anastomosing complex of melt veins and channels concentrating upward into a main mush column stalk with periodic sills (Marsh, 1996; Korenaga & Kelemen, 1998). The axial magma chamber beneath the ridge axis caps the mush column.

The system works through hydrostatic head produced in response to, in effect, suction at the top associated with the abrupt splitting and spreading of the lithospheric plates. From an initial state of slight over-pressuring or near hydrostatic equilibrium, this abrupt motion splits the brittle oceanic crust, withdrawing magma from the underlying axial magma chamber, some of which erupts as pillow lavas, and what is left in the feeder solidifies as a dike. The loss of hydrostatic equilibrium at the head of the system propagates as a pressure pulse downward throughout the system, perhaps partly as solitons, drawing magma upward in the mush column and re-establishing stability. Because all parts of the contiguous ridge plates do not move at exactly the same time and to the same degree, the process of melt motion at depth is certain to be complicated. Magma at some depths may sometimes be transported laterally as it ascends, making its overall trajectory significantly non-vertical.

The intimate time-series history of this process is recorded in the vast sheeted dike complex, which is a unique characteristic by-product of ridge magmatism well known from ophiolites. Because these dikes are random temporal samples of the axial magma chamber, they give throughout time an excellent inventory of the general state of this magma, both in terms of composition and crystallinity. These dikes are typically fine-grained and of low phenocryst content; any phenocrysts are mostly plagioclase. The axial magma chamber or sill is a passive body that experiences bursts of withdrawals followed by recharging from below. Some recharges undoubtedly carry massive amounts of entrained large crystals, principally olivine, which sediment to the floor, undergoing punctuated differentiation immediately upon entering the sill. Layering is thus an indirect reflection of dike formation.

Mid-Ocean Ridge Basalts (MORBS) erupting at the ocean ridges worldwide are tholeiitic basalts, with low $K_2O$ (~0.2 wt.%), modest $TiO_2$ (~1.5 wt.%), and low and un-fractionated REE (rare earth elements). To first order these basalts are chemically among the most globally uniform of any class of magmas, reflecting the underlying magmatic process of prolonged intimate contact with an extensive mass of solids in the magmatic mush column. In effect, the chemical composition is buffered by the mechanics of the overall process, much as in a household water purification system. This uniformity also reflects the uniformity of mantle composition itself, which has probably not changed drastically over Earth history. That is, at the present rate of ridge magmatism ($\sim$20 km$^3$/year), with about 10 percent melting the mantle will be recycled about once every 5 Ga, which, even allowing for the role of contamination by subduction, suggests the mantle composition has been, to first order, fairly constant throughout Earth history. Due to variations in spreading rate along with slight variations in mantle composition, melting depth and bulk composition cause second order chemical variations (e.g., Klein & Langmuir, 1987).

It is interesting to consider ridge magmatism from the past classical point of view of magma chambers, where primitive (MgO-rich) magma is generated at depth, emplaced in a large vat, and differentiates by crystal growth and settling, generating a long "liquid line of descent" and a large chemical diversity of products from picrites to olivine basalts and tholeiites, and possibly even granites. Instead, there is exceedingly little diversity found at ocean ridges. Even picrites (olivine-rich basalt) are not found at ridges, and arguments have always persisted that if picrites are parental to MORB then why are they never seen? Why don't they erupt more often, or at all? This is not an uncommon situation in petrologic studies where critically important parts of the hypothesized dynamic puzzle are not seen but are nevertheless postulated to be in a hidden zone or are un-eruptible due to density difficulties or other factors. On the other hand, the fact that critically important magmatic parts of the system are never seen may simply mean that they as actual physical processes don't, per se, exist as postulated. Instead, a more reasonable process achieves the desired chemical effect, like uniformity. In the present understanding of ridge magmatism, the magmas are chemically fractionated by contact with olivine, and other phases, through prolonged intimate, diffusional, contact within the mush column, not classical crystal fractionation. The lack of presence of an actual massive, 6–8 km thick magma chamber, as postulated on chemical grounds, is replaced by a virtual magma chamber, which is an integrated chemical process taking place over the full extent of the mush column.

This, again, emphasizes the value of being able to interrogate an extensive magmatic system, like the Ferrar system of the McMurdo Dry Valleys, in its natural setting. It is the intimate linking of petrographical and chemical variations with spatial variations that is highly unusual.

## 1.4 Magmatic Environments

A striking difference between these two magmatic systems is the regional thermal regimes in which they exist. Beneath ocean ridges the entire system resides within a massive thermal upwelling, with high temperature country rock extending outward to great distances; all magmas are, in effect, thermally insulated and can exist in a perpetual steady state almost regardless of size and shape. It is only when magma is evicted from the high-level sill as a dike when severe cooling and quenching takes place.

This is in strong contrast to magmatic mush columns existing in the lithospheric-crustal regimes where the containing country rock is much cooler than the magmatic system (Figure 1.1). Cooling is rapid from all boundaries and there is a constant, if ephemeral, intense competition between active magmatism and progressive solidification. For the system to stay alive for any significant period of time, chilled magma must be periodically replaced by fresh hot magma. There is a wide spectrum of longevity or states of thermal survival measured by the volumetric throughput or ongoing flux of magma through these systems. Large systems, like Hawaii, thermally dominate the lithosphere and can exist for millions of years, leaving a massive thermal imprint on the adjoining lithosphere. Small systems, like those associated with monogenetic volcanism, may only evade thermal death for a thousand years or so. Magmatism associated with Gondwana rifting begins as small weak systems sometimes die when rifting is aborted, and with successful rifting become massive ocean ridge systems. Perhaps the most important, most telling, and valuable systems are those that become strong and well developed but then, due apparently to tectonic manifestations, fade. Yet, besides establishing a mature plumbing system, they impose an intense thermal imprint on the crust, sometimes leading to extensive remelting and reworking of the entire adjacent crust. It is these systems that reveal much of the intimate workings of mature magmatic systems. This is what is exhibited in the McMurdo Dry Valleys.

# 2

# The McMurdo Dry Valleys Magmatic System
# (Ferrar McDV)

## 2.1 Introduction

At the center of the history of the development of the Transantarctic Mountain belt (Figure 2.1) is the tectonic history of Antarctica itself, which, of course, came about as a separate entity via the breakup and dispersal of the supercontinent Gondwana some ~200 Ma ago. Much has been explored and discussed about this history, especially what precipitated the breakup, but it still remains somewhat murky (e.g., Storey, 1995; Stump, 1995; Faure & Mensing, 2010, p. 497, etc.). The earlier assembly and later dispersal each point to Antarctica as a core continent wrapped in a collection of all the southern continents, plus India (Elliot, 1975). Besides the actual dispersal phase, which gave rise to the Ferrar magmatic system, the earlier history of assembly, with possibly prolonged subduction along the later rifted margin, enters squarely in understanding the subsequent rapid development of the magmatic system, taking perhaps only ~1 Ma, and the characteristic compositions of the sills and lavas, including the great similarity of these to those in neighboring continents. The net result is a mountain range providing a special environment for understanding basaltic magmatism within continents, and its role in reshaping continents themselves.

Direct observation of magmatic systems through field studies is pretty much limited to continents and by the natural vertical range of Earth's topography, which for eroded igneous terrains is, practically speaking, about 4–5 km. Geophysical studies have great global and vertical range but are limited in spatial resolution. The most reasonable approach is thus to combine indirect information from geophysical studies of large active systems like Hawaii (e.g., Ryan, 1987, 1988, 1993) and the mid-ocean ridges (e.g., Sinton & Detrick, 1992) with direct mapping of deeply eroded, well exposed crustal basaltic systems (e.g., Hayes et al., 2015).

The Ferrar McDV contains information of unprecedented vitality to this approach. The heart of the system is an expansive (~10,000 km$^2$) stack of four

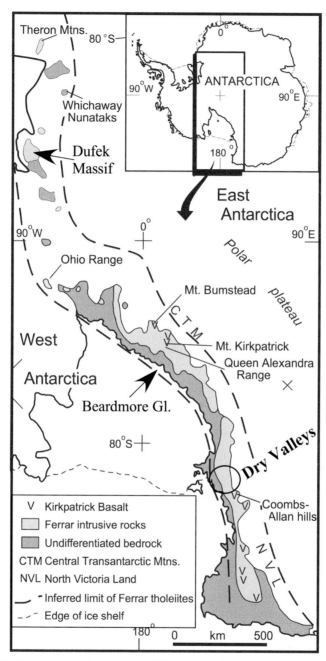

Figure 2.1 The general distribution of Ferrar basaltic rocks, both volcanic and intrusive, in the Transantarctic Mountains; note the positions of the Beardmore Glacier region, and the Dufek Massif relative to the Dry Valleys (after Elliot & Fleming, 2017).

massive sills, each ~300–500 m thick, centrally interconnected, and exceptionally well exposed over a vertical extent of about 4 km, culminating in the direct connection to a regional flood basalt, the Kirkpatrick Basalt (e.g., Elliot & Fleming, 2017, 2021). The volume of intrusive magma involved alone may have been 10,000 $km^3$ or more. These rocks record the episodic or serial establishment, loading, and eruption of a magmatic system with compositionally and temporally contiguous aliquots of compositionally diverse magma. Magma composition varies systematically upwards through the system from ultramafic (20% MgO) orthopyroxenite at the base to tholeiitic (basaltic andesites, with 4–5% MgO) in the uppermost sills and lavas. The overall bulk chemical compositions describe a strongly differentiated and diverse system exceedingly similar in overall range of bulk composition to that defined by the lavas of Hawaii (i.e., Kilauea, Mauna Kea, and Mauna Loa; Figure 1.1). And, as at Hawaii, MgO content is directly related to the presence of magnesian phenocrysts, orthopyroxene in the Ferrar McDV, and olivine at Hawaii (e.g., Murata & Richter, 1966; Garcia et al., 1995, 2003). In striking contrast to Hawaii's lavas, however, the spatial context of every sample defining the differentiation sequence within the Ferrar McDV magmatic system is accurately known. This spatial control allows the processes giving rise to the compositional variations to be ascertained at an unusual level of detail, making the connection between plutonism and volcanism unusually clear.

Moreover, the lowermost sill, the Basement Sill, contains a massive tongue of these large (1–20 mm) orthopyroxenes, the abundance and size of which varies systematically throughout the sill, also extending vertically upward, partially filling the overlying Peneplain Sill. The presence of these crystals acts as a tracer from which the dynamics of emplacement, differentiation, and sorting has been recorded. This high concentration of large orthopyroxenes, for example, acted during shear as a sieve to the much smaller plagioclase, pervasively forming anorthositic stringers or schlieren and layers exclusively in this Opx-Tongue. These features are clear local indicators of flow direction, including the local dynamics of fractional crystallization and differentiation. Moreover, local ponding in the topographically deepest part of the system has formed a small (~500+ m thick), but exceedingly well-defined, layered ultramafic body exhibiting many of the sorting and sedimentary features of Earth's large ultramafic, layered intrusions, like Stillwater. The relatively modest thickness of this body and the sills in general allowed rapid cooling (~1,000 years per sill), thus preserving intimately diagnostic textural relations that are commonly lost to annealing and fully erased in larger layered bodies. The preserved textures are exceptionally clear and record unusual examples of annealing and grain growth through annexation of adjacent small grains through grain boundary migration.

A corresponding isotopic contrast between the Basaltic Carrier Magma and the Opx-Tongue materials, indicating entrainment from a subcrustal/upper mantle ultramafic regime dominated, not by olivine, but orthopyroxene, compounds this petrologic diversity. This intimate coupling in primitive phenocryst/primocryst nature and content within a voluminous distinctive basaltic magma makes this system singularly valuable to understanding the fundamental transition between Gondwana rift-related magmatism and eventual ocean ridge establishment.

The core of this work is to describe the important field relations and petrology of this system in some detail. Special emphasis is placed on the exact spatial context of a vast suite of chemically analyzed samples, with the aim of revealing the basic geometric development, dynamics of magma transport, and mechanics of differentiation within the system. This is followed by an effort to relate this to the larger picture of the ultimate magma generation and ascent leading to development of the greater Ferrar magmatic province.

I start with the history of discovery and early previous work and then describe the overall structure, field relations, sampling strategy, internal sill stratigraphy, petrography, rock chemistry, and style of functioning of the overall system, with a final emphasis on its relevance to the general dynamics of development of dolerite Gondwana rift systems, magmatic systems in general, and the possible ongoing communication between continental crust and the uppermost mantle though crustal delamination processes. As will become evident, this region intimately displays some of the most detailed and important magmatic geology on Earth.

The history of discovery of this region is rather poignant on many singular levels, and the intimate blend between bold exploration, raw survival, and scientific discovery is essential to appreciate in some detail. It is only through a detailed account that the exceptional efforts of these seminal efforts can be fathomed. They built the foundation of all later research and, in some real way, this spirit of science-based exploration still exists in Antarctica to this day.

## 2.2 Discovery

### 2.2.1 James Clark Ross (1841)

Little did James Clark Ross (1800–1862) know that, after discovering the exact spot of the North Magnetic Pole on Boothia Peninsula near King William Island in the Arctic Spring of 1831, in his quest to similarly locate the South Magnetic Pole, he would on 27 January 1841 traverse with his ships HMS *Terror* and *Erebus* a wide belt of pack ice and sail into a beautiful open Sound containing a huge active volcano (Figure 2.2). Although the Magnetic Pole was nearby, but too far inland to be reached, he made note of the majestic ice clad mountain range to the west, then

Figure 2.2 James Clark Ross ships in McMurdo Sound, January 1841, with Mt. Erebus in the distance (from Ross, 1847, vol I, facing p. 217)

named the Sound for Archibald McMurdo of HMS *Terror* and the volcanic island for his good friend and mentor Sir John Franklin, governor of Tasmania. He collected some rocks, including a wide assortment of igneous pebbles from penguin and seal stomachs (Ross, 1847; Prior, 1899), and turned and headed back North to Portsmouth, arriving in 1843 to much acclaim and a knighthood.[1] This was the first true sighting of the Transantarctic Mountains (TAM, Figures 2.1 and 2.3), an imposing range stretching some 4,000 km across Antarctica from North Victoria Land on the Ross Sea to Coats Land on the Weddell Sea.

Discovery of the vast repository of beautifully exposed dolerites came about by accident as early explorers, seeking the South Pole, attempted to cross this range to gain access to the inner Polar Plateau containing the Pole itself. The first of these expeditions to again enter McMurdo Sound was Robert Falcon Scott's *Discovery* expedition (1901–1904). The path to discovery of the deeply hidden McMurdo Dry Valleys was arduous, tortuous, not at all obvious, and scintillating.

---

[1] These two ships, *Erebus* and *Terror*, were refitted, complete with thickened ice-hardened hulls and steam engines, and put under the command of Sir John Franklin to seek and traverse the much-coveted Northwest Passage; leaving in 1845 with 128 men, Sir John and his men were never again seen alive by the western world. Commissioned in 1812, HMS *Terror* shelled Ft. McHenry in Baltimore in 1813, inspiring Francis Scott Key's poem "The Star-Spangled Banner."

Figure 2.3 The Royal Society Range of the Transantarctic Mountain Range, as viewed across McMurdo Sound from Hut Point, near McMurdo Station. From this common uninviting view all along this formidable mountain front, it is little wonder that the easily accessible Dry Valleys went undiscovered until Albert Armitage attempted to scale the mountains directly in 1902.

## 2.2.2 *R. F. Scott's* Discovery *Expedition (1901–1904)*

After struggling through the Antarctic outer pack ice, following in the footsteps of James Ross, Scott was surprised to enter the beautiful open water of the Ross Sea. The white continent was on the horizon. On Wednesday, 8 January 1902, they arrived first at Cape Adare, some 350 km Northwest of where they would eventually settle and visited Carsten Borchgrevink's Hut. On board was Louis Bernacchi, a physicist, who had wintered over with Borchgrevink (1864–1934) in 1899 as part of the Southern Cross British Antarctic Expedition of 1898–1900. This was the first such party to winter over on the actual continent at Cape Adare adjoining Robertson Bay. One of Scott's five civilian scientists among the full party of 50 men, Bernacchi's experience now became invaluable in pointing out landmarks and navigating the coastline further South, for Borchgrevink had cruised deeply into and made a cursory survey of McMurdo Sound itself, including Ross Island. Cape Adare presented a good example of an awkward base camp, a cramped spot up against cliffs, offering relatively little opportunity for exploring the hinterland. After two days looking around, making magnetic measurements, and getting their first feel underfoot of Antarctica, at three in the afternoon they weighed anchor, carefully plying southward along the coast looking all the time for possible spots to winter over. Scott was anxious, the season was getting on, a home base to spend the winter had to be found.

Progress was slow as northward-moving streams of ice forced *Discovery* away from the shore, often making close inspections difficult. But they persevered as strong headwinds slowed them to a standstill on 12 January in approaching Coulman Island, named by Ross for his future father-in-law. Ice-filled Lady Newes Bay and Terra Nova Bay came next, and then *Discovery* skirted the impressive Drygalski ice tongue extending well out from the shore, an unusual impressive glacial feature they would come to know well. Pressing on to Cape Gauss they found the Nordenskjold ice tongue and were eager to see Wood Bay coming up, which had been discussed at length in England as a possible base. But the general nature of the coastline already told them that they would be trapped here against ice-bound mountains, just as Borchgrevink had been at Cape Adare.

Scott realized he needed to get closer to shore and explore a little on foot at the next likely snug harbor, which they found on the afternoon of Monday, 20 January. He nosed *Discovery* in as close as possible, but the shore ice was fast and a large shore party was "… soon bounding from floe to floe, now and then encountering a breach too wide to be leaped and having to raft themselves across." Large boulders of "beautiful crystalline" granite on the shore with streams of clear freshwater winding through the stones amid banks of bright green luxuriant moss gave this attractive inviting inlet the name of Granite Harbor. And the way the inlet hooked around to the north gave "… a promise of snugness and security about this spot which we met nowhere else." As inviting as this spot was, the best seen so far, with the season getting on there was a growing urgency to put in roots, but here again a direct route to the south would be at the whims of sea-ice. Scott thought this might serve as home if nothing better could be found, so after a long day just after midnight *Discovery* turned southeast into McMurdo Sound. Pursuing a base at Granite Harbor would have been a major mistake as the fast ice, firmly welded to the shore, seldom leaves here, and if it had been later in the season and *Discovery* had been berthed here it is very likely she would still be there today.

In pulling away Scott scanned the shore with his strongest binoculars hoping for another possible spot; nothing appeared and in clear sight 200 km directly to the east stood Mt. Erebus. In "gloriously fine" weather, "By 8 A.M. we were in the middle of McMurdo Sound, creeping slowly, very slowly, through the pack-ice, which appeared from the crow's nest to extend indefinitely ahead." Twelve hours of slow going in the pack ice took *Discovery* further south, south of Earth's South Magnetic Pole, and in the far distance at what seemed to be the ice-bound head of McMurdo Sound a distinctive conical mountain appeared, taking the honorable name of their sturdy ship, Mt. Discovery. The South was firmly blocked by the massive ice shelf.

So at 8 P.M. on the 21st we thought we knew as much of this region as our heavy expenditure of coal in the pack-ice would justify us in finding out, and as before us lay the great unsolved problem of the barrier [Ross Ice Shelf] and of what lay beyond it, we turned our course with the cry of Eastward ho!

Heading east along the edge of the Ross Ice Shelf, all they found was a massive thick sheet of ice, with cliffs in places over a kilometer high. Finding a place to get up, they set up a hydrogen air balloon and Scott went up to look around. Dropping too much ballast he was nearly carried away, but he also saw that there was nothing here but a vast, seemingly endless sheet of ice. To be sure, Lt. Albert Armitage, second in command, was sent inland with a sledging party to judge the surface and scan the horizon; nothing promising. Scott decided to give McMurdo another look, perhaps there might be a spot further south of Granite Harbor, which had been ice choked three weeks ago. With the temperature dropping worries set in that the season was closing in about them and perhaps they might come up empty handed or be caught and forced to winter over in a bad spot. A disaster could be in the making. Things were tense.

On the evening of Friday, 7 February, while approaching McMurdo Sound, snowstorm after snowstorm hit *Discovery*. Not being able to see beyond the bows, they stopped, furled the sails, and brought up the steam. The temperature started coming up with a southeast wind and the skies began to clear. By Saturday morning they were pleasantly out in the Sound again in much better air, finding much of the ice had vanished. They made hard for the western shore, near the mountains, perhaps there was yet something good.

. . . as we eagerly scanned the coast of the mainland our hopes rose high that we should find some sheltered nook in this far south region in which the "Discovery" might safely brave the rigours of the coming winter, and remain securely embedded whilst our sledge parties, already beyond the limits of the known, strove to solve the mysteries of the vast new world which would then lie on every side.

They headed directly for rocky cliffs south of Granite Harbor and again ran into formidable ice. Between the close headlands of this New Harbor a massive glacier came down from the inland mountains,[2] but only a narrow possibility existed of gaining shelter behind an ice tongue should the pack ice clear. As the sky began to clear, open water appeared to the southeast. They headed for it. The dreary clouds lifted and on the sunny, glistening Saturday afternoon of 8 February they sailed unusually deep into McMurdo Sound, feeling special, men crowding the rails as they plowed through thin, 3–4 inch, night ice. No one had ever been this far south. It was a special year. The ice tide was turning in their favor. Vast high mountains of South Victoria Land on the starboard to the west, cloaked in brilliant ice, running south as far as they could see (Figure 2.4). Ross Island, with its active, smoking volcano Mt. Erebus just in front, was off the port bow. In the far distance, Mt. Discovery becoming ever more prominent, looking every bit like another volcano. The cards were down, a good base was needed. It was perhaps now or never, there had to be something hereabouts. Next came Scott's greatest discovery: Hut Point.

---

[2] This would later be called the Ferrar Glacier and the headlands to the north the Kukri Hills. Armitage later led a sledge party up the mountains to the south and realized that this was a major corridor to the inland Polar Plateau. Scott would follow with his own sledging party up the Ferrar Glacier.

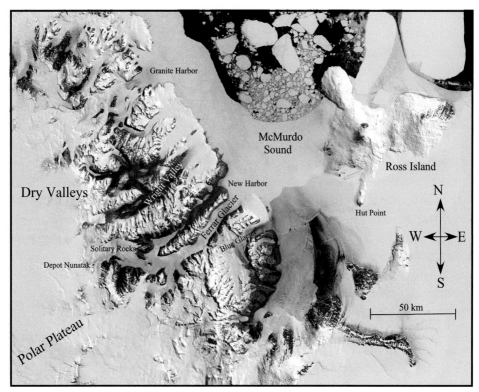

Figure 2.4 The physiography of the Polar Plateau, McMurdo Dry Valleys, McMurdo Sound, and Mt. Erebus on Ross Island
(adapted form Landsat Image of Antarctica (LIMA); http://lima.usgs.gov)

The open water led them closer and closer to the southern extremity of Ross Island. The fast-oncoming Autumn seemed delayed here, the edge to the wind was softer, night ice was hardly forming. There was more time to look around before worrying about being accidentally frozen in. Now Mt. Terror and the southern-most slopes of Erebus were in sight. Suddenly the presence in front of a low final wall of fast ice said that this was the end of the road south, at least by ship. There was no special southern strait. Yet beyond the last tip of Ross Island stretched the broad flat plain of the Barrier connecting everything between them and Mt. Discovery and several other low ridges.

Carefully edging along the ice edge brought *Discovery* northward again, close in alongside a steep cinder-laden spur forming a slight bay in the edge of a long narrow peninsula (Figure 2.5). At 10 P.M. Scott closely read the water using a hand lead line and then inadvertently finally nosed forward until *Discovery* gently grounded herself on a bank a few yards offshore. (Almost exactly two years later, in attempting to leave, this seemingly gentle bank nearly destroyed *Discovery* and

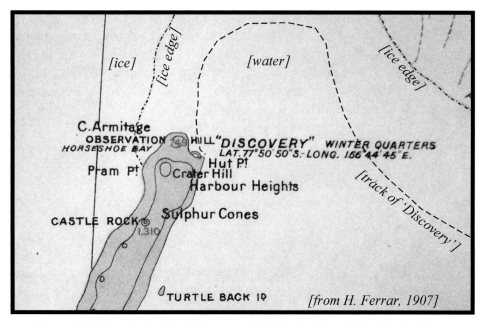

Figure 2.5 Map of the Hut Point area, McMurdo Sound, by Hartley Ferrar indicating the ice edge, the track of *Discovery* and its wintering over location, and Observation Hill, among other features (after Ferrar, 1907)

her crew.) Needing something secure to anchor to, Scott backed off and moved a few hundred yards north around this spur, coming in against foot ice where he used ice anchors to secure the ship for the night. It had been a long and anxious day. But here, right in front of them, Scott recognized an ideal location for their base. It hadn't come too soon. He was excited and pleased.

We have now the possibility of making this part of the bay our winter quarters. From the point of view of traveling, no part could be more seemingly excellent; to the S.S.E. as far as the eye can reach, all is smooth and even, and indeed everything points to a continuation of the Great Barrier in this direction. We should be within easy distance for exploration of the mainland, and apparently should have little difficulty in effecting a land communication with our post office at Cape Crozier. There are no signs of pressure in the ice; on the other hand, the shelter from the wind is but meagre, and one can anticipate intense cold and howling gales. On the whole, to-night I feel like staying where we are. *(Scott, 1905, v. I, p. 210)*

First thing next morning Scott and Armitage went ashore and walked south a short way. Every step was on some product of volcanism: cinders, ash, lava, and volcanic bombs. With no sign whatsoever of animal or plant life, they found, in effect, a truly lunar landscape. Just to the south where *Discovery* had first grounded they overlooked a small snug ice-filled bay. Incoming swells had broken the ice and, in a few days, they figured it might be open enough to give *Discovery* a fine winter berth. Various parties roamed the hills and ice and, upon reconvening in the evening there was a rush of excited conversation over all that had been seen. This was clearly home; names were coming forth. The southern-most point is Cape Armitage and the high cone Observation Hill.[3] The next volcanic ridge to the north became Crater Heights and the highest point Crater Hill with the valley in between here and Observation Hill becoming the Gap. The first little bay they tied up in just to the north of Hut Point became Arrival Bay and the final cove to freeze in became Winter Quarter Bay. In the near background to the east was Arrival Heights and further away to the northeast a higher steep-sided butte became Castle Rock, which would be a constant navigational sentinel for crossing the peninsula. For now, excited by their good luck, they were eager to start settling in.

In a very real way: Their Eagle had landed at Tranquility Base.

From Australia, they brought a prefabricated wooden Hut with each stout wall panel labeled by a small, coded aluminum marker, to be erected on the near slopes. Only under necessity would they live here. It was much too small, being 36 feet on a side with a large overhanging eve, to house this big crew. They would stick with their trusted quarters on board *Discovery*. The Hut, which Scott ultimately saw as dispensable, he also prophetically saw as a haven to ". . . afford the necessaries of life to any less fortunate party who may follow in our footsteps . . ." It became indispensable to all McMurdo Expeditions, but now it was a place for work, science, and play. A magnetic observatory, to monitor Earth's magnetic field, in the form of two small huts, was put up out behind. In spite of the reluctance of the ice to leave the little bay and of still having difficulties getting a good secure spot for *Discovery*, they got to work with vigor off-loading the materials for the huts and began to erect them in earnest. Each morning breaking through the night ice with the whaler, landing a crew at the ice foot.

The importance of finding this location, clearly the single best spot for over a thousand miles of coastline and recognizing it for all its high-quality features cannot be overemphasized. Seeing the obvious is not easy. Undoubtedly Scott's greatest discovery, a splendid mastery of seamanship under trying conditions. It is easy to see why he would feel so parental to this location, as he spent much of the last 10 years of his life in this vicinity. He would know every nook and cranny and would die 100 miles south of here returning from the Pole.

---

[3] It was here that they would go to look for the return of the deep field sledging Southern Parties, and where at the summit would be placed in 1913 the jarrow wood cross commemorating the loss of Scott and his men returning from the Pole, containing Alfred Lord Tennyson's last lines from his *Ulysses*: "To Strive, To Seek, To Find, and not to Yield." It was blown down in the winter of 1992, repaired at Scott Base, laboriously and carefully carried, and reestablished by a number of us in February 1993.

This was, and is, a very special jeweled spot on a hostile edge of Earth, Scott's spot, and it had not been easy to find. It quickly became the Jerusalem of Antarctic exploration. Scott later made Ernest Shackleton, who was with him in 1901, promise in writing that he would not go here to make his attempt on the Pole in 1907. And when Shackleton, like Scott himself, could find nowhere else to go, he had to come back here, although the ice was not with him and he, nor anyone since, could ever be so lucky as to get so deep into McMurdo Sound again without an icebreaker. Shackleton thus fell deeply from the good graces of many, splitting allegiances between "Scott Men" and "Shackleton Men." But Shackleton should have known better. He had seen the coast. He too had been up in Scott's hydrogen balloon. He knew what to expect. There was simply no other piece of solid ground so accommodating for exploration.

The original plan back in England called for a small wintering-over party to be left here to live in the Hut and for *Discovery* to return north to New Zealand. But with the discovery of this fine location, Scott felt confident that no harm would come to the ship by being frozen in here. This was a serious decision for it meant that he would have a large group of novice and eager explorers on his hands for a whole year, through a long tough dark winter. With no outside contact possible whatsoever once into winter, Scott was taking a big chance that he knew what he was doing, and he had all the necessary supplies for 50 men to survive a full Antarctic winter (Figure 2.6).

The stage was now set. Once *Discovery* was securely anchored and laced to the shore, which turned out to be difficult, and then frozen in, the sea voyage came to an end. All ensuing activities would be land-based, needing men with the expertise of making serious assaults on polar terrains. But there was no one here of this ilk. Armitage, Koettlitz, and Bernacchi had some experience, but this was limited. The rest were rank amateurs. On board *Discovery* they were experts at seamanship, trained by the best navy in the world right down to understanding the slightest nuances of handling all aspects of a ship under all conditions. And up to now this expertise showed in all that they did. There were no disasters; all functioned smoothly. On land, however, they were ripe for one disaster after another, and the first one was only days away. In a series of life-threatening learning experiences, they learned sledging, camping, using primus stoves, using sleeping bags, staying warm, how to walk, ski, shovel, and survive in snow. It was not pretty.

Scott himself was not able to participate for several weeks as he had seriously wrenched his right knee in a fall on skis and stayed on board in a splint. Trips were made first to White Island by Ernest Shackleton, Edward "Bill" Wilson, and Hartley Ferrar, and then Cape Crozier by a large group, to leave a message on their whereabouts, and finally Scott lead a depot-laying trip south using dogs. After three days trying to figure out how to haul the sledges, put up the tents, get into

Figure 2.6 The Hut at Hut Point in 1902, with *Discovery* in the background and the magnetic observatory in the foreground (Scott, 1905)

their sleeping bags, cook, and cajole and drag the dogs, they had covered 6.5 miles due south. Scott decided it was fruitless to go on and cached all the supplies, around 2,000 lbs., and high tailed it for home. They made the trip back, with the dogs going great guns, in six hours. It was good fun, but all realized the formidable task ahead before they could become accomplished, successful sledgers. One seaman, George T. Vince (1880–1902), died on the Cape Crozier trip; when returning to Hut Point, he became disoriented in a blizzard and fell into the Sound. They had a long winter to hone their skills and get ready for a full campaign of Spring and Summer sledging.

### 2.2.2.1 Albert Armitage's Discovery of the Dry Valleys

In the Spring (September 1902) 14 sledging trips (Figure 2.7) were planned and carried out, the most important of which was the one by Scott, Wilson, and Shackleton directly South over the Ross Ice Shelf and along the mountains in the hope of finding a path to the South Pole, which was the main reason for the entire expedition. While Scott was away, Albert Armitage, second in command, took on the challenge of going directly inland to the west. He had already been to the west side of McMurdo Sound, to the edge of the mountains at New Harbor, in September, when scurvy had broken out in his crew. He found the Ferrar Glacier,

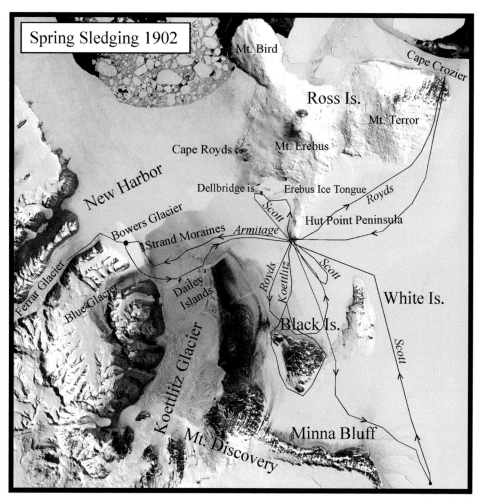

Figure 2.7 The planned *Discovery* sledging trips for Spring (October–December) 1902, each emanating from Hut Point

at least at the foot, much too rough to ascend. But he had seen enough to be hopeful that once beyond the lower reaches it might smooth out sufficiently to allow deep access into the interior. Although appearing massive and impregnable,[4] Armitage now put together an extensive sledging trip into the Western Mountains. This was to be a journey of epic proportions in the discovery of the McMurdo Dry Valleys. "It was a fine, clear morning on November 29, 1902, when the teams who were to endeavour to wrest their secrets from the mighty mountains which faced us in the west were drawn abreast the ship" (Armitage, 1905, p. 158).

---

[4] It is recorded, at above an apocryphal level, that more than one modern scientist having arrived in McMurdo, in looking across at these mountains, similarly decided not to tempt fate in such terrain and returned home.

Scott had all along been skeptical of this enterprise, but Armitage was persistent, and Scott's skepticism was not ungrounded as the view from Hut Pint across to the lofty mountains (Royal Society Range) shows a massive apparently impenetrable, formidable terrain encased in ice (Figure 2.3). A perfectly impossible task for Armitage. He reiterated this in his final instructions to Armitage: "In regard to the western exploration, I leave the detail unreservedly in your hands. I fully approve of your plans and of the men you propose to select, and shall expect you to take such aid as you require for extending your trip" (Armitage, 1905, p. 144).

Armitage put together a well-planned, well-equipped, and well-peopled expedition. His approach was markedly distinct from that of Scott. In consisting of 21 men, it was the largest sledge party of the whole expedition. He arranged the group into four teams, A, B, C, and D; teams A and B, with six men each, as the "Main Party" had provisions for eight weeks and teams C and D, with five and four men, respectively, as the "Supplementary Party," had provisions for three weeks. Everyone was equipped with skis and ski poles and enough sledges to keep the pulling weights to about 225 lbs. per man for each team. Every detail of food and equipment was carefully gone over by Armitage and discussed with the others. This was to be a trip of exploration and science. One smaller sledge of the A team held the instruments for measuring the inclination, declination, and overall intensity of the magnetic field, which would give first-hand information on the present position of the South Magnetic Pole, which was about 350 miles northwest of them. This sledge was easily separated from the rest of the Main Party for Armitage and Skelton to make measurements while the others maintained the pace. The Supplementary Party was a geologic party, with Koettlitz, as both physician and geologist, in charge overall and a member of C Team, and geologist Hartley Ferrar heading up the D Team. The Team members were:

A: Armitage, Skelton, Scott, Buckridge, Quartley,[5] Evans;
B: Allen, Macfarlane, Wild, Walker, Hendsley, Duncan;
C: Koettlitz, Croucher, Clarke, Whitfield, Dellbridge;
D: Ferrar, Dell, Pillbeam, Hubert.

They left *Discovery* on Saturday, 29 November, and headed due west–northwest across the Sound for the "Moraines" found on their earlier trip, camping at the south end of these features among boulders, gravel, and large blocks of ice, a craggy strange collection of ice and rock. Here they dispatched two seals to augment their diet, and Skelton demonstrated his new expertise at snaring Skuas, which they found delectable. On the following day began a long, difficult, and

---

[5] Wilson (1966, p. 337) says that Quartley, a stoker, was an American and that he had brought along a little black "she-cat" that lasted until it succumbed to the sled dogs on 3 February 1904.

slow ascent up a glacial valley (Blue Glacier) into the mountains. The plan was to circumvent the rough ice on the Ferrar Glacier by hitting it much further upstream. This route would not be easy. Armitage expected as much and had come prepared with block and tackle, crowbar, and ice axes. Driving the axes in deep to hold the crowbar and block, the sledges were hauled up steep inclines, and coupled with all working together to relay the sledges forward, in two days they ascended over 3,000 feet. Here Armitage himself went forward on skis to seek a path over to the Ferrar Glacier. After about a half mile he "came to a sharp ridge, lined with crevasses, and a glorious scene suddenly came opened into view" (Armitage, 1905, p. 165). The pass to the glacier some 2,000 feet below looked much too steep for the sledges and more hauling brought them to an upland plateau from which he soon saw there was no way for further ascent. They were trapped and a wave of despair passed over Armitage as he realized all this work might well have been in vain. Having now been away for almost two weeks, here Koettlitz's party of Teams C and D got set to return. Koettlitz examined Armitage's party and pronounced them fit for further exploration, but also saw in most of them incipient signs of scurvy and cautioned Armitage to be vigilant about the possibility of an outbreak.

The only possible way ahead might be to the north; going ahead with seaman Scott and Wild they came upon a small, isolated ridge formed of two granite mounds (Granite Knolls) with a saddle in between. Off to the northwest the snow field "descended precipitously to an undulating plain of ice" (p. 169) filled with crevasses and holes. Here again there was no way ahead for sledges. Back at camp bad weather came in, pinning the party down for three days. When the weather finally broke the sky and heads cleared and the nearby descent to the undulating plain of ice now seemed doable and sensible. Here they went, allowing the sledges to travel under their own heads and at the foot of the slope camp was made.

First thing in the morning, Skelton, Evans, Wild, and Armitage examined the long, steep pass ahead through which the sledges would have to travel to reach the Ferrar Glacier. Sitting on an empty sledge with ropes running individually to him and the sledge, the others lowered Armitage down the first slope. As down he went, he sounded with his ice axe to scout for gaping holes and crevasses that might swallow men and sledges whole. At the end of the rope, the sledge was anchored, the men came down and the performance was repeated until a descent of 650 feet had been accomplished. Working ahead from one down-slope to the next over a distance of several miles northwestward the Ferrar Glacier below came within reach. Thus, Descent Pass received its appropriate name (Figure 2.8): "The whole pass formed a magnificent amphitheatre of ice-rock and snow about half a mile in breadth at its widest part, and three miles in length" (Armitage, 1905, p. 171).

Figure 2.8 Scenic view of Descent Pass, looking down on the Ferrar Glacier and the Kukri Hills

Returning to camp they found the sledges all loaded and ready for the ordeal. Dragging them to the summit of Descent Pass, the sledges were lashed two abreast with ropes under the runners for brakes and ropes out bridal-style behind for further braking. Four men accompanied each pair and, for the first, Skelton, Allan, Macfarlane, and Armitage had the privilege. "It was the most exhilarating run, far more exciting even than the water-chute at Earl's Court. We all arrived safely at the bottom of the first slope, and the remainder was comparatively easy" (Armitage, 1905, p. 171).

All down safe and camped on the Ferrar Glacier at the foot of Descent Pass, sea mist rolled in and held everyone in place for another day before any move could be made (Figure 2.9). Far ahead the pathway seemed blocked by large mountains, but there was no telling how far this frozen highway would lead them. Here they unknowingly sat on the most direct and beautiful pathway to the Polar Plateau, a ribbon of ice leading them through the most ancient relict terrain on Earth. But that was all in the days ahead, for the immediate days they had to learn how to move on this slick ice and, if this wasn't bad enough, a fresh fall of four inches of wet snow left the surface even slipperier on foot and more obscure. It was a good thing they had fashioned heavy-duty crampons from the broken windmill, brought to generate electricity at the Hut, but a failure in these winds.

Figure 2.9 Ferrar Glacier, looking west toward the Polar Plateau, with Cathedral Rocks to the South (left) and the Kukri Hills to the north (left)

Once on the move, Armitage went ahead on skis, picking a route and feeling for and marking crevasses. With the fresh sticky snow clogging the runners, friction high, and having to relay sledges, progress was slow, yet it was exhilarating to be in this striking mountain corridor lined by steep bare mountain ridges on each side. Hugging the smoother ice formed by seasonal meltwater on the left or south side, about six miles were covered by the end of the day. Up ahead in the far distance the mountains blocking the way now seemed to perhaps be showing a small opening on the right that might let them pass, but it would be another day or two yet before they could be certain. A fine campsite was found against a series of sharp cliffs carved into pinnacles and buttresses looking ". . . at a little distance like some vast cathedral falling to decay" (p. 172). These forever became Cathedral Rocks and nearby a cache of supplies was made on the lowest ledges (Figure 2.10). Looking about near here in the evening they came upon carcasses of crab-eating seals, freeze-dried and mummified, some 35 miles from McMurdo Sound and about 2,000 feet above sea level.[6]

Held down again by weather for 36 hours progress remained slow until after ascending waves of ice ". . . what looked like frozen rapids" (p. 173), rougher but cleaner ice was found, progress improved, relaying ended as all sledges were hooked together and again hauled in series; their preferred mode of transport. Now the mountains blocking the way showed their true stuff: they did not block the way

---

[6] Mummified seals of this nature have been found throughout McMurdo Dry Valleys. Some carcasses may be as old as 5,000 years. One found in 1994 by the author at Solitary Rocks had, during the Autumn of 1993, ascended some 65 miles up the Ferrar and Taylor Glaciers, entered and traversed Simmons Valley on bare rock and sand, climbed 1,500 feet up a steep rough, rocky slope, died, and was carved by winter sandblasting into a mammalian ventifact. They are generally not old seals and, of the more than several hypotheses on this behavior, are thought to do this because of simple curiosity or plain simplemindedness.

Figure 2.10 Armitage's camp on Christmas day 1902 among the moraine of the
Ferrar Glacier, with Terra Cotta Mountain in the background (Scott, 1905, v. II,
facing p. 138)

but bordered a large glacial basin. By Wednesday, Christmas Eve, camp was
made among a series of large rock blocks carried northerly in a medial moraine
by merging glaciers just north of a distinctive double knob-topped mountain:
Knob Head.

   Here were 12 solitary men cautiously entering an almost mysterious mountain
inner sanctum that in the 50 million years since formed by erosion had never seen
life beyond the occasional intruding deranged seal and sightseeing skua. There
were no permanent residents, not even bacteria.[7] This was not just any place on
Earth. This was a singular Place on Earth. An exceptionally special place on Earth.
Here was an unexpectedly beautiful ice basin, sitting deep within a majestic,
massive, and towering mountain range. A natural train station for glaciers: glaciers
arriving and departing at all points. As evening came on and the splendor of the
setting enveloped and diffused into their senses, a lengthy discussion ensued over
the rock formations and the major glacial contributions. Ice flows coming in from
the south and west, merging to form the Ferrar, which itself split, forming north
and south arms. The flow from the south came down from the high inland
mountains, appearing uninviting as a route to the deep interior. The north arm after
a few miles seemed to turn around the sharp nose of the peninsula-like series of
hills (Kukri Hills), forming the north wall of the corridor they had been following,
and then heading back toward the Sound. But the western route, although perhaps
rough with icefalls in sight, seemed an obvious direct route to the high inland Polar

---

[7] Nematodes at 50 microns, cyanobacteria in glacial ponds, and scarce lichens would eventually be found.

Plateau. This was all an exciting discovery, especially the absolutely unexpected snow-less mountains and brilliant colors of the rocks forming remarkably horizontal strata.

> The mountains to the south of us, bordering the glacier, were quite bare of ice and snow, and their colouring was most beautiful, consisting of all shades of red, brown, and yellow. The layers of light (sandstone) and dark coloured (basaltic) rocks ran in remarkably even lines throughout the range, except where here and there a fault occurred. *(Armitage, 1905, p. 176)*

Why were the mountains so free of ice and snow? How could this be, so near Earth's South Pole; the world's best museum of ice and snow? This would take some time to digest. No immediate answers were obvious. They were wide-eyed.

After being held down by challenging winds for another 36 hours, stretching out Christmas Day, the 12 men dragged the sledge train onward and upward over rippled ice with the weather warm enough to collect pools of water in the hollows, furnishing refreshing drinks and making the surface almost greasy in slipperiness. True icefalls began appearing at an elevation of about 4,000 feet, separated by a mile or so of rippled ice. These were laboriously negotiated under constant fear of dropping through into bottomless ice caverns.

In the distance as each rise was gained what could be seen of the horizon beckoned to them as an endless expanse of snow and ice. Armitage and Skelton were especially excited by the prospect of reaching the final interior where they might see, once and for all, the true personality of the Antarctic continent. Names came forth as the basin was crossed. An island-like mass of rock off to the right, layered as clean as a cake, became Solitary Rocks (Figure 2.11). And, to the south, west of Knob Head, became Beacon Heights. But the sweetest, sharp peninsular finger of land, with the spectacular striking sheer face of basaltic rock caught in the

Figure 2.11 Upper Glacier Basin from the southwest end of Kukri Hills, looking westerly toward the Polar Plateau

Figure 2.12 Finger Mountain area in foreground and Turnagain Valley on the left, with dolerite sill within Beacon sediments

Figure 2.13 Finger Mountain, with sills within Beacon sediments

act of invasion and eating sandstone, bending out and pointing due east, became Finger Mountain (Figures 2.12 and 2.13). The deep turning majestic, *cul-de-sac* valley behind with high headlands southward in the heart of the mountain range eventually became Turnagain Valley. Features so grand that they would become singular examples of Earth's beauty and inner dynamics.

Moving ahead the ice dipped down gradually before it rose again into a broad basin. Still slippery and sharp, the sledge runners suffered greatly, nearly stripping the German silver away, and the wood fared no better. But the pathway onward to the west remained wide open; all the anticipated mountainous obstacles fell by the wayside. In the distance the slopes began again with chaotic ice *rapids* marking the boundaries between the steps. With attainment of each rise the horizon invitingly stretched out before them until, finally, on the last day of 1902, after 25 miles, the mountains fell behind and a vast expanse of flat, endless, majestic snow and ice began to unfold to the west; something on a far grander scale that only Nansen and his Norse companions had seen some of on Greenland's Plateau. But here the scale was unimaginable. Once up at this elevation, about 6,500 feet, the ice grew smoother but now large sweeping crevasses came into the picture. Roofed over with crusted snow and barely discernable except by the ridges at the edges; Armitage and later Scott himself with Evans would in due time fall into and become intimate with these crevasses.

The last vestige of any approachable rock loomed up in the distance as a dark mass of highly jointed basalt, a distinctive glacial nunatak, providing a fine location for a high-level depot of a week's supply of food and an extra sledge. This became Depot Nunatak, beyond which the ice became an amphitheater headed by an 800-foot-high balcony of rapids; just as in a river where oncoming water bunches up to get enough force to get around a massive boulder (Figure 2.14). This was the dear price to pay for the shelter of the rock, being the summit of a buried mountain. The sheets of black magmatic rock intruded within thick slices of bright sandstone throughout the region brought forth the realization that the icefalls also reflected exactly this structure. The black magmatic rocks being highly resistant to erosion would not easily give way to the ice, damming and ponding it and producing frozen waterfalls, impeding sledging progress and making the sledgers think.

Figure 2.14 Depot Nunatak, made up of columnar jointed dolerite, on the edge of the Polar Plateau, found by Armitage's party on 31 December 1902 (Scott, 1905, v. II, facing page 140)

Cautiously the party moved ahead, rising steadily each day until on the New Year reaching an elevation of 7,500 feet. While moving smoothly at a temperature of –2°F and a strong W–SW wind, Armitage stopped, turned, and looked back to the struggling Team B and was startled to see the sudden collapse of Macfarlane. Rushing back, he found Macfarlane's breathing ragged with chest pains and an ashen face. After his experience in October with scurvy, Armitage was alarmed; a tent was put up and warm fluid poured into him. He seemed a little better, at least stable. Being right on the verge of making a deep traverse onto the Polar Plateau, Armitage and Skelton were reluctant to turn back. After some consultation and observing that Macfarlane seemed stable and might be suffering from mountain sickness and not scurvy, Armitage decided to have Team B make camp here while Team A went on for at least a few more days. The Polar Plateau was just too much of a sterling discovery, and rightly so, for them to turn back now. With no officer to accompany them Team B could not be sent back to lower elevations, but with Frank Wild available they were in good hands, even though the crew said Wild's heart beat like a locomotive at night. So, Armitage, Skelton, Evans, Scott, Buckridge, and Quartley pushed on for another five days, covering 25 miles in a W–SW direction moving on what seemed a limitless expanse of whiteness.

There was not much to see, just long swells in the ice and in the far distance to the S–SW two more black nunataks rising above the ice. But, just the same, every stride made history; landing on and exploring a new planet. Time would show nothing else like it on Earth. On Sunday, 4 January 1903, their last outbound camp was set; the highest elevation attained being 15 feet shy of 9,000. The way to the Polar Plateau from McMurdo had been unlocked; the Polar summit had been attained, right through the heart of what all along appeared to be an unassailable massive wall of mountains. Just to see what was beneath their feet, while Armitage took magnetic measurements of the total field intensity, inclination, and declination, the others dug a hole four feet wide and seven feet deep. No ice was found, only layer after layer of snow recording the seasons over the past 20 years. At the bottom a three-foot ice axe pushed downward still found no ice. So, this is how glaciers form came the thought. The next day, striking out from camp, another five miles west was covered, sightings taken, and everyone stared at the nothingness. A limitless expanse of whiteness filled all eyes. In the far distance to the southwest a large whaleback ridge of mountains stuck up, to become the Lashly Mountains. Back at camp, amidst much conversation over the surround-ings, gear was tidied up for the retreat. In the morning, Tuesday the 6th, camp was struck, and they hurried back along their tracks toward Depot Nunatak, Armitage worried all the time about the fate of Macfarlane.

As the six men moved across the white expanse, in the afternoon sun the high tan and red sandstone cliffs in the flat topped, butte-like mountains to the north stood out like sentinels guiding them back to harbor. These mountains became the Inland Forts, and the bright beckoning sandstone became the Beacon Sandstone, a formation that in five years Shackleton would also find as far away as the Beardmore Glacier over 400 miles south. The single most characteristic formation of Antarctica. By late afternoon, with a strong tailing wind, Allen's camp came into view in the distance and two speck-like figures, who even from a distance looked like Allen and Walker, soon started moving toward Team A. Entering the first of the complicated crevasse fields before Depot Nunatak, Armitage jokingly yelled over to Scott not to fall into just to celebrate his birthday today. The words were barely out before Armitage himself dropped 27 feet (Scott says 12 ft.) downward into a large crevasse. A pinnacle at 17 feet had conveniently slowed his descent. Hanging by his sledging harness, the ice walls sloping away in each direction and continuing downward into blue darkness, Armitage soon heard Skelton's reassuring voice and felt a rope land on his head. In quick order he was out, shaken, but no worse for wear. They all hustled on to Allen's camp where the new arrivals stuffed themselves into a single tent for a hearty hot meal, while the tent walls heard the news of reaching the summit of the Plateau. All men were healthy except Macfarlane who was still weak and had to travel aboard a sledge upon departure next morning, Wednesday, 7 January 1903.

At Depot Nunatak the week worth of provisions was retrieved and the stored sledge cannibalized for parts for the other sledges.[8] Progress was rapid. The way was known and now the countryside could be inspected a little more closely. A campsite was made "... near a magnificent-looking perpendicular cliff which showed most markedly the different formations of rock" (Armitage, p. 185). They were back to Finger Mountain. After breakfast the next morning Skelton and Armitage went to the face, descended a steep gully along the foot and found good rock samples in the loose fallen rocks. They also found here a thin child-sized, sheet-like intrusion (a sill) of basaltic magma penetrating evenly and perfectly horizontal for hundreds of meters. How magma could penetrate this fast and so far without congealing was puzzling. On the way up they had passed Finger Mountain at a greater distance and now the full scale of this remarkable face was even more appreciated (Figure 2.15). Skelton, the ever-persistent photographer, did his job well and never put off the occasion to record a scene. Hartley Ferrar was highly pleased to see these samples and the photographs. On this evidence he later came with his own party (Ferrar, Weller, Kennar) along with Captain Scott to make a detailed examination of the whole area while Scott went as far as he could on the Plateau.

---

[8] Remains of this depot and the one left later by Scott the following season were casually looked for at various times from 1993 to 2008, with no success.

Figure 2.15 Armitage's party at the face of Finger Mountain, with a thin dolerite sill intruding Beacon sediments (Scott, 1905, v. II, facing p. 140)

Presently the two teams again headed rapidly down the glacier, making a beeline for the foot of Descent Pass some 35 miles east. With some slipping, sliding, and falling at the icefalls, the trek went smoothly, passing on their right deep valleys, dry and remarkably free of ice, cutting for long distances into the heart of the mountains, the Royal Society Range. Taken with the splendor, Armitage and Skelton could not help but give names to the sandstone heights, Beacon Heights, and some of the mountains, Pyramid, New Mountain, Terra Cotta, and Tabular. From the shapes of the nearest slopes, an ice-line or bathtub ring could be discerned, indicating that the ice level had once in past times been 2,000–3,000 feet higher. Perhaps the climate had warmed significantly.

With the approach to Descent Pass temperatures rose to 40°F, producing large amounts of melt water, encountering shallow ponds as large as a mile long and quarter mile wide, and allowing the men to bathe their feet and drink their fill. Over the ensuing winter this water froze to provide long expanses of smooth ice for easier travel by Scott's and Ferrar's parties in the Spring. Once at the foot of Descent Pass, Armitage evidently thought only of returning the way he had come, which meant dragging everything back up the steep Pass. The first part was by man hauling, but soon block and tackle was needed. Twelve hours of hard labor on the 12th put the whole crew at the top of the Pass. Now it was all more or less downhill. Macfarlane still did not do well; sometimes traveling on skis behind the sledges he did okay, but then repeatedly fainted and had to be hauled along. The Moraines came into sight and were reached on the 17th and fresh seal meat made for a "regal spread," but the skuas were now too wily. Across the Sound the familiar landscape about Winter Quarters beckoned them and at 7:30 P.M. on Monday, 19 January, after being away for 52 days, Armitage's teams arrived back at *Discovery*.

All was running smoothly at *Discovery*. Lt. Charles Royds had the ship ready for departure as soon as all were back, and the ice went away. Scott was still away on his southern journey with Wilson and Shackleton, now struggling home, arriving on 3 February, all with strong signs of scurvy, perhaps Shackleton the worst. Ferrar and Koettlitz were away gauging the effects of glaciation at Minna Bluff. And Barne was about to return from his six-week trip southwest to the foothills.

The relief ship *Morning* had arrived at the ice edge, some 10 miles away on 24 January under the assumption of escorting *Discovery* back to New Zealand, where pending sufficient funds it might make a second trip South to McMurdo. Assuming the ice in the Sound would break up and be blown away, as the month went on concerns heightened when the ice was reluctant to leave, even with the encouragement of explosives. Finally, with the season getting on and new ice beginning to form, and the ice edge still some three miles away, Scott decided, rather than abandon *Discovery*, the best course was to winter over again. Supplies were shifted across the ice, and some men were given the chance to return North, which slimmed down the cast a little. Ostensibly worried about his health, Scott also ordered Shackleton north, much to Shackleton's everlasting disappointment, and igniting in him the burning desire to return with his own expedition, which he did in 1907. He also encouraged Armitage to go north, as his wife was on the verge of birthing a child when they left, but he didn't bite. Staying over another winter gave an excellent opportunity for more detailed examinations of the Dry Valleys.

### 2.2.2.2 R. F. Scott's Exploration of the Dry Valleys

Scott's reaction to Armitage's major discovery of a path through the Royal Society Range was, understandably, measured joy. Although not a daring attempt on the Pole, time has shown this to be a far more fundamental discovery, even providing a route, albeit awkward, to the Pole itself. Scott assessed what Armitage had done and explained why further exploration was needed:

It was evident that this party had reached the inland ice-cap and could claim to be the first to set foot on the interior of Victoria Land; but it was clear, too, that they had been forced to terminate their advance at an extremely interesting point, and to return without being able to supply very definite information with regard to the ice-cap. As I have already pointed out, the view of the sledge traveler on a plain is limited to an horizon of three or four miles; beyond this he cannot say definitely what occurs. This party appeared to have been on a lofty plateau, the very short advance they had been able to make over it could not give a clear indication of what might lie to the westward; the nature of the interior of this great country was therefore still wrapped in mystery. *(Scott, 1905, v. II, p. 144)*

Like Queen Victoria, it was hard for Scott to be generous. A skillful and interesting writer, he possessed the rare ability in writing to walk a fine edge in appearing modest, parentally insightful, and condescending all at the same time. In sum: Yes, Armitage had done something perhaps interesting and important, but the true value was still uncertain until vetted by Scott himself. And from the ice swells traversed by Armitage and Skelton, they certainly could see more than 3 or 4 miles; the nunataks observed to the S–SW were at least 30 miles away. Still, to stand on any margin of the Polar Plateau gives an instant impression of a majestic infinite whiteness. Yet, Scott also immediately realized that the geology of this newly discovered region was critically important to examine in some detail, especially since sandstones (*quartz-grits*) had been found that might contain fossils. Ferrar had run out of things to do, and, as good a place as Hut Peninsula was for a base, it is technically not a part of the Antarctic continent, and so far, while with Armitage, Ferrar had had only a tiny taste of the continent. For all intents, Winter Quarters, just as McMurdo is today, may well have been a sequestered penal colony. No one so far had seen anything up close of the true nature of the continent. All that anyone had seen, prior to Armitage's discovery, were the rocks around Granite Harbor visited on the way down in *Discovery*. And, try as he might, Scott had been unable to reach rock on any of his sledging trips. So here finally was a first-class job for Hartley Ferrar and something for the whole expedition to focus on for the upcoming season. "On the whole, therefore, the western party had done excellent pioneer work; they had fulfilled their main object, and in doing so had disclosed problems which caused the greater part of our interest to be focused in this direction throughout the remainder of our stay in the South" (Scott, 1905, v. II, p. 145).

**Laying Cathedral Rocks Depot**   In anticipating the long major sledging trip, Scott lead a Spring trip to the Ferrar Glacier to lay a depot and to seek a more direct route to the Dry Valleys rather than suffering the arduous climb to Descent Pass. He took a party of five men, Skelton, Dailey, Evans, Lashly, and Handsley, and left on Wednesday, 9 September. Due to bad weather and lack of familiarity with the approach, five days elapsed before camp was made near the foot of the Ferrar Glacier, which is a jumbled mass as it meets the ice of the Sound. With the season still young, temperatures dropped to –49°F, and the nearby hills blocked the low Sun, giving only cool deceptive twilight by which to navigate. Scott was worried; nothing looked easy. The following day, Tuesday 15, cheery sunshine made the day and Scott and Skelton set out on foot to try to pick a way through the rough and tumble ice (Figure 2.16). The others climbed the nearby bare hillside to the south-west to get an overview of the area. By taking time to wander back and forth trying to stay on smoother ice, Scott and Skelton found themselves being directed more toward the center and north, right hand side, of the valley and made fairly good headway. Back at camp, the scouting party also reported apparently smoother ice in

Figure 2.16 Rough ice at the foot of the Ferrar Glacier, looking southwest toward Cathedral Rocks

the distance to the north and upstream. It was now or never, and they wasted no time in immediately beginning to portage all the gear, including sledges, over the rough ice.

In the far distance a blue ribbon of ice containing dark streaks of boulders marked a possible good route. With each mile the ice became more manageable and by late Wednesday afternoon the lower reaches of Descent Pass were nearby. Continuing ahead in the morning on the smooth ice made of last year's melt water, they were soon abreast of Cathedral Rocks. The supplies were cached near a large conspicuous boulder, bearings taken, and then all turned to head home. Scott was pleased with himself. It had taken him about 10 days to get as far as Armitage did in 3 weeks. On the downward journey, by steering even closer to the north edge the ice was found to be even smoother. They were struck by the well-known feature of glaciers to be smoother where the Sun does the least damage, which is on the north sides of east–west glaciers in high southern latitudes. They were learning, and Scott was enthused to think they would have quick and easy access to this route without further sledge portaging. He had put his foot in the door and could claim some bragging rights, which was always so important to him and to most expedition leaders. "We were inclined to be exceedingly self-satisfied; we had accomplished our object with unexpected ease, we had done a record march, and we had endured record temperatures . . ." (Scott, 1905, v. II, p. 209).

*Discovery* was reached by the 22nd, where they found that Barne, taking over eight days in laying a depot near White Island, had felt even lower temperatures, down to –67.7°F when the thermometer broke.

Telling everyone of the new route, the path and plan was clear. Scott had fretted over the winter about being unable to get the ship free. But now he looked upon this detention in the ice as an "unmixed blessing." And that by exploring this new region the ". . . the value of our labors of the first year would be immensely increased" (Scott, 1905, v. II, p. 219). This had to be a major comforting realization; a whole new playground for exploring and science was at hand. And, even though he had seen virtually nothing of what Armitage had found, ". . . there were fascinating problems elsewhere, but none now which could compare with those of the western land. It was such considerations that made me resolve to go in this direction myself, and I determined that no effort should be spared to ensure success" (v. II, p. 219). Moreover, this territory was right in McMurdo's backyard, a veritable playground for science and exploring.

The many parallels between this trip and Scott's last march to the Pole in 1911–1912 manifest themselves in Scott's actions and words. Whether he knew it or not, he was entering into a pattern that he would repeat, a profound déjà vu. And his actions seem to have brought forth in him exceedingly similar emotions, which he described in similar words. The deep parallel beginning here was marked by following a pathway discovered by someone else, Armitage here, Shackleton later, who had found, cut, and named the trail, Ferrar Glacier here, Beardmore Glacier later, named the most distinctive landmarks, and in each instance only had to go a little farther, about a hundred miles in each, and be a little more dramatic to make a name for himself. Scott had the lifelong knack of never himself really discovering anything significant on foot. The singular exception being Taylor Valley on this trip. He almost always went where others had already blazed the trail. This came from his trait, common among many, of thinking that the biggest most sensational discoveries come from the hardest and longest efforts over the greatest distances, which is only true when the full lay of the land is known. On this trip his great strength of being a naturalist explorer would come to light in spades.

Scott began by laying out a template of words of how great this journey was planned and performed, tending toward hyperbole and speaking as if no one had been there before:

Rarely, I think, has more time and attention been devoted to the preparation of a sledge journey than was given to this one. I rightly guessed that in many respects it was going to be the hardest task we had yet undertaken, but I knew also that our experience was now a thing that could be counted upon, and that it would take a good deal to stop a party of our determined, experienced sledge travelers.

I am bound to confess that I have some pride in this journey. We met with immense difficulties, such as would have brought us hopelessly to grief in the previous year, yet now as veterans we steered through them with success; and when all circumstances are considered, the extreme severity of the climate and the obstacles that stood in our path, I cannot but believe we came near the limit of possible performance. *(Scott, 1905, v. II, pp. 219–220)*

He was, indeed, mainly correct about him, particularly him, and his men with regard to their fitness and stamina, but much less correct in his attention to fundamentally serious details on the equipment side at all levels.

**The Main Event**  Scott's "Western Party," consisting of three groups, set out for New Harbor on Monday, 12 October. His party contained: Scott, Skelton, Feather, Lashly, Evans, and Handsley. Skelton and Evans had been with Armitage out on the Polar Plateau. Ferrar's supporting and accompanying geologic party contained: Ferrar, Kennar, and Weller. And Dailey's supporting party contained: Dailey, Williamson, and Plumley. Twenty minutes after they left, Skelton hurried back to fetch the hypsometric thermometers left behind. And a short time later, the home crew found the traveling theodolite and tripod, an essential item for mapping, had also been left behind. Royds rushed out, caught up, and delivered the much-needed instrument just as they discovered its absence. After sitting around all winter, Scott wanted to show this group of men new to sledging with the Captain what a real sledging journey was all about:

As I had determined that from the first to last of this trip there should be hard marching, we stretched across the forty-five miles to New Harbour at a good round pace, and by working long hours succeeded in reaching the snow – cape on the near side early on the 14th – a highly credible performance with such heavy loads. *(Scott, 1905, v. II, p. 221)*

Nevertheless, on the 14th things seemed to start going a bit downhill:

Had to camp early to-night as Dailey and Williamson are a bit seedy, probably a little overcome with the march. At supper the third member of this unlucky unit, cut off the top of his thumb in trying to chop up frozen pemmican. He is quite cheerful about it, and has been showing the frozen detached piece of thumb to everyone else as an interesting curio. *(Scott, 1905, v. II, p. 222)*

In just four days out they reached the Cathedral Rocks Depot, loaded up the supplies, and headed another few miles inland to a row of stakes Armitage had placed the previous year to gauge the flow of the glacier. Here camp was made among large boulders of the medial moraine where enough snow had drifted to secure the tents. By Saturday the 17th Armitage's old campsite at Knob Head Moraine was reached. Scott had made in 6 days what took Armitage 27. Scott was pleased, and he was especially pleased with the sights and the grandeur of the world about him. His poetic naturalist side began to burst forth:

The changes of scene throughout the day have been bewildering. Not one half-hour of our march has passed without some new feature bursting upon our astonished gaze. Certainly those who saw this valley last year did not exaggerate its grandeur – indeed, it would be impossible to do so. It is wonderfully beautiful. *(Scott, 1905, v. II, p. 225)*

But for the main part we are surrounded with steep, bare hillsides of fantastic and beautiful forms and of great variety in colour. The groundwork of the colour-scheme is russet brown, but to the west especially it has gradations of shade, passing from bright red to dull grey, whilst here and there, and generally in banded form, occurs an almost vivid yellow. The whole forms a glorious combination of autumn tints, and few forests in their autumnal raiment could outvie it. *(Scott, 1905, v. II, p. 226)*

Not only is this highly accurate, but it purveys the emotional sense one gets in traveling from the volcanic cinder-laden black and white world of Hut Point Peninsula to the Polar Shangri-La of the Dry Valleys. Scott's feelings for the interplay of color and landscape are highly developed and, in this sense, it is no surprise that he found an artist with these same sensibilities to be his best friend, Bill Wilson, and later his Impressionistic-artist wife Kathleen Bruce, a friend of Isadora Duncan and Henri Matisse.

The lovely sanctity of the environs came to a crashing halt when on this very evening, the 17th, after achieving all this mileage, the carpenter, Dailey, reported the German silver covering the runners had split on two of the four sledges. The rough sharp ice of Ferrar's Glacier had made quick work of them. And for being so early in the trip, it was certain that the others would soon be in a similar shape. With these heavy loads, stripping the metal off and running on bare wood might soon render the sledges useless. Armitage had experienced similar trouble, but his warnings had apparently not been taken seriously. Scott thought now with the ice getting smoother they might perhaps manage until reaching snow higher up, but he was worried. Moving on past Mt. Hendsley, Knob Head, and suffering the numbing katabatic blasts from Windy Gully, camp was made opposite Solitary Rocks, closer to New Mountain. At suppertime on the 18th Scott thought it best to unload all the sledges and make a thorough inspection "... with horrid revelations." Only one sledge had decent metal, the others useless. This was a real emergency. The entire sortie was in danger of ending right here. The only possible recourse was to depot all the gear, take a half-week's supply of food with the three bad sledges, and do a rapid forced march back to the ship for thorough repairs. This they did.

Scott was bound and determined to lose as little time as necessary and to show his team, and Ferrar's also, what real hardened sledger-men could accomplish. Now 87 miles from the ship, they headed off early on the 19th and made it to the ship in three days, arriving in the evening of Wednesday the 21st, covering 37 miles on the last day. Ferrar's team also returned, arriving 2 hours after Scott, having covered over 38 miles, setting a mileage record for the whole expedition.

Five days were spent repairing the runners, double or sandwiched runners were employed, giving added insurance against sharp ice. During this time Scott realized that his original extremely well-organized sledge journey would have to be completely reorganized. The original 12-man group was reduced to 9 men: Dailey, Williamson, and Plumley, who had cut off the end of his thumb, would stay home. Ferrar's geological group would be on their own, disengaged with a single small 7 ft. sledge working outward from the depot opposite Solitary Rocks. Ferrar chose to stick with Scott as long as possible and then go off on geological work while Scott pushed out on the Polar Plateau. Scott's "advance party" and a single 11 ft. sledge would fetch the left behind sledge and supplies cached off Solitary Rocks and "... dispose of them on a new plan." Now lightly loaded the sledges were much less vulnerable to the ice and on Monday morning, 26 October, the caravan set off again. By having to repeatedly negotiate the foot of the Ferrar Glacier, the best possible path was found and in three days camp was made opposite Descent Pass. Nevertheless, again, during light hauling on the 28th, on the first icy incline, the under runners split and had to be removed. Scott was exceedingly annoyed and was "... determined to get to the top this time, even if we have to carry our loads." For the remainder of the trip "... we had constant worries with these wretched runners." On the following day Ferrar's sledge collapsed and a long delay was suffered in making repairs. Yet, on the 29th camp was made within a few hundred yards of Armitage's Knob Head Moraine campsite. So far, all told, Scott had expended 17 days to get here as compared with Armitage's 27. More surprises were just around the corner.

Sledging along in the large ice basin of the Dry Valleys, the winds calmed, and the temperatures rose, lulling the men into almost pastoral comfort. Clothing, sleeping bags, finneskoes, socks, and hats lay about the tent at breakfast time drying in the gentle warm Friday morning sun of 30 October. Wind always blows here and no or little wind means simply that the wind is changing directions and will commence shortly with gusto. Sure enough, caught lounging about with breakfast in hand strong gusts swept down on them, scattering gear everywhere.

The incident would have been extremely funny had it not involved the possibility of serious consequences. The sleeping bags were well on their way towards the steep fall of the north arm before they were recovered, and by good luck the whole affair closed with the loss of only a few of the lighter articles. *(Scott, 1905, v. II, p. 236)*

In no time a full-force gale was up and running, forcing camp to be reset to the south over the glacier edge in the lee; persisting for two days, no progress was made. Pinned down in the gorge between the glacier and the Knob Head headland, there was ample time to study the details of the ice, its stratigraphy, and constituent sediment load (Figure 2.17). Scott took particular interest, showing his keen eye for

Figure 2.17 Edge of Ferrar Glacier, near Knob Head, showing included sediment, see man for scale: Scott's party on 30 October 1903 (Scott, 1905, v. II, facing p. 237)

detail. Careful observations showed the gas bubbles (vesicles) and sediment to be restricted to certain levels, and the darkest central layer had no sediment or vesicles, the purest ice yet seen. On the nearby upland slopes, large fields thickly strewn with glacial boulders told again of past times when the climate was colder and the ice much more extensive.

The first morning of November proved quiet enough to move on, past two more desiccated seal carcasses, past Windy Gully ("Vale of Winds"), and on to the depot made 12 days earlier opposite Solitary Rocks. At first all looked intact, but the lid of the instrument box had not been secured enough, had blown open, and some of the contents lost to the wind. Skelton's goggles and "one or two other trifles" were gone, but "... I found to my horror that the 'Hints to Travellers' had vanished." From the title a seemingly harmless loss, yet this publication put out by the Royal Geographical Society contained all the necessary navigational information and calculation aides for determining longitude and latitude. Without it out on the Polar Plateau Scott would never quite know where he was.

When this book was lost, therefore, the reader will see how we were placed; if we did not return to the ship to make good our loss, we should be obliged to take the risk of marching away into the unknown without exactly knowing where we were or how to get back. *(Scott, 1905, v. II, p. 240)*

Another trip back to the ship would have killed both the time schedule and the sledges, putting the whole excursion in jeopardy. Scott talked it over with the men. He was loath to return yet again to the ship and, after thinking about the problem for 10 days, he figured he could make do for an estimate of latitude with a simple measurement of Sun angle. He did this by making a free hand graph in his journal of the Sun's declination over time for certain known latitudes. At Winter Quarters he had recorded in his journal the Sun's first appearance of the year and of the day when it first remained above the horizon at midnight. And he had observations on this trip at known latitudes. Connecting these points with a straight line and extending it gave enough information to estimate latitude, which turned out to be quite accurate. Nothing similar could be done for longitude, which involves more extensive calculations, and these would have to remain until back at the ship. If he had been more of a land-based man, Scott might have navigated more directly simply by compass and mileage. Armitage had already been here and measured the full magnetic field intensity and declination and inclination. But even this was not needed, just a good compass and a mileage meter would do.

Onward and upward the whole crew continued, giving a wide berth to the lower and upper dangerous icefalls of majestic Finger Mountain in whose north wall is an entire geological curriculum. Swinging out wide to the north above Solitary Rocks brought them near the high walls of the mountains holding the Inland Forts and, with high gusty winds coming on and Ferrar's runners needing repairs, they took harbor in the horseshoe cove near Beehive Mtn (Figure 2.18). As the men worked on the repairs, Scott looked around at the massive shear walls all about and was much impressed by the evenness and "perfect uniformity" of the beautiful sandstone strata.

**Upward to the Plateau**　In strong southerly winds Scott now steered south to get in the lee, which brought them into calmer air but chaotic ice, learning the hard lesson of glacial travel to avoid close approaches to land masses, especially points and peninsulas. All the rest of the day was spent finding a way through this ice, tearing up the runners. But the vistas were striking in clean, clear, warmer air (Figure 2.14). Now at 7,000 feet many of the more stratospheric mountains down in the valley to the east appeared smaller, with summits beneath them and the Royal Society Range itself giving way to more manageable terrains and isolated islands of dark rock. Proceeding upward and southerly, on Wednesday, 4 November, they entered the basin at Depot Nunatak, admiring the fine "columnar

Figure 2.18 Beehive Cove area (on left) of Inland Fortes, northwest of Solitary Rock, where Scott's party camped on 2 November 1903. This is the Asgard Sill in the Beacon sediments.

Figure 2.19 Upper Taylor Glacier, looking southwest toward the Polar Plateau with the crevassed ice-bowl in the foreground, near Depot Nunatak on the far-right hand side

basalt" and taking shelter from the wind gusts rolling down from the Plateau. Ahead was the bow wave of ice approaching Depot Nunatak and once beckoned into this basin the only way out is up a steep, crevassed slope of blue ice. Winds off the Plateau cascade into this one-sided bowl-like basin, meeting at the bottom, beating and buffeting any and all below (Figure 2.19). Little did they know this would be a miserable home for the next week.

With the weather fast turning to a gloomy overcast, a rapid chill set in at −24°F and a plateau ground blizzard hit them halfway up the slope, which was much longer than expected. Everyone was fast becoming frostbitten in the face, and Scott frantically sought a safe place for shelter with a snow patch to secure the tents.

They trundled back and forth, eventually finding a patch of dense old snow. Scott seized a shovel himself to help make camp but couldn't make a dent. Finally, they settled in and, after brewing tea and coiling up in sleeping bags buffeted by flapping tent canvas, Scott felt relieved to "... have come mighty well out of a very tight place. Nothing but experience saved us from disaster today ..." Here at "Desolation Camp" they were pinned down for a week, and as is universally felt by travelers under similar circumstances, Scott was, again, no different: "... and if I were asked to name the most miserable week I have ever spent, I should certainly fix on this one." This was an exposed forward position and the only possible way through it was to hunker down and wait it out, hoping all the time that the tents would survive and none of the men would do something dangerous. Twenty-two hours of each day were spent buried in sleeping bags, the wind howling outside beating the tent wall back and forth as if someone is violently trying to get in. The height of the rattle of pellets of ice and snow on the tent walls reflect the intensity of the gale. The lulls bring startling silences and fresh hope of finally a welcome end to the imprisonment. But just as spirits rise and heads come up to expose an ear to listen more carefully, it hits again with even more vengeance; beating all again into deeper submission and showing who's boss.

The remaining two hours in each day were spent over a hot meal in the morning and evening; the bags were rolled up and as much as possible a decent meal fixed. Some ongoing state of normalcy being essential to preserve what little remains of well-being. Conversations center on when the weather will break, and a close vigil is maintained by a meteorological buff on the barometer to gauge the regional system with an eye to incoming high pressure and cheery blue skies. There seems to have been little in this group in the way of weather prognostication, but they did have one book in Scott's tent: "Darwin's delightful 'Cruise of the "Beagle".'" They took turns reading until fingers "... refused to turn the pages." Conversations between the tents was only possible in lulls, but the sledges were each hauled under a tent and the metal stripped from the runners in preparation of soon greeting snow. Sleeping bags progressively iced up and a general sense of cold malaise began to creep through the camp. Seriously cold and frostbitten hands and feet began to appear. Scott realized regardless of the weather they had to escape. They had to get up and move on, and he had to be the one to make the move.

Before breakfast on Tuesday, 10 November, Scott, Feather, and Skelton got into marching footgear and began digging out the sledges and gear. The wind was low, but after breakfast down came the gusts again and, in attempting to complete digging out, Scotts hands and Skelton's toes and heel on one foot became frostbitten and Feather lost all feeling in both feet. A yell over to the others amidst lulls gave similar reports and Scott, seeing futility in their efforts, called for retreat. The wind, he felt, was even worse today; maybe tomorrow would be better.

Scott was beginning to doubt the whole exercise. The next morning, they recommenced digging out, hurried through breakfast, bundled up the sledges and, fearing new gusts at any moment, made a dash for the top of the basin.

Ferrar had had enough, having used up much of his precious time and supplies hanging on with Scott and with no more rock in sight. He had already seen a fair amount, but there was much more to see and sample. After hurried instructions from Scott, he turned with his crew back to Depot Nunatak for shelter and samples. He and his tiny band of Kennar, and Weller, with their 7-foot sledge, trekked back and forth in the enchanting ice basin below visiting all the cliff faces and outcrops they could possibly get a hand on. His geologic map and extensive report (Ferrar, 1907, 1925), for a freshly minted undergraduate geologist with almost no experience and much limited time, is remarkably good.

Scott's party was now down to six men on two 11-foot sledges. His party contained Evans and Feather, and the other had Skelton, Lashly, and Handsley. As soon as they reached the rim of the bowl, having carelessly scrambled up slope over bad crevasses and around gaping chasms, they suddenly emerged into low winds and good visibility. And now they were almost exactly on the same schedule as Armitage. Over the next two days they moved up and down over long swells and finally arrived at 8,900 feet whereupon the "summit" had been achieved. The immaculate white Polar Plateau stretched endlessly out before them (Figure 2.20). In clear skies, Scott was deeply impressed by the sights of the Royal Society Range to the east, behind them. They had enough food for five more weeks of travel, which meant about two more weeks outbound.

The only course to take, to better what Armitage had already done, was to head due west out on and into the whiteness, to go beyond the horizon Temperatures now dropped each day, getting down to –44°F, which gave little chance to dry out

Figure 2.20 The Polar Plateau, looking southwest, which Scott reached on 20 November 1903

sleeping bags, iced-up in Desolation Camp. And the true nature of the Plateau set in with winds of 10–15 mph relentlessly in their teeth all day long. Not a big wind, but a cold annoying and tiring wind. Yet the conditions and the sanctity of the environs came with spiritual rewards.

I don't think that it would be possible to conceive a more cheerless prospect than that which faced us at this time, when on this lofty, desolate plateau we turned our backs on the last mountain peak that could remind us of habitable lands. Yet before us lay the unknown. What fascination lies in that word. Could anyone wonder that we determined to push on, be the outlook ever so comfortless. *(Scott, 1905, v. II, p. 254)*

**Into the Unknown**    With the other team struggling relative to Scott's team, at lunchtime on Sunday, 22 November, the teams were reorganized with Feather going to Skelton's group and Bill Lashly coming with Scott and Evans. Scott had another week of travel in front of them before they had to turn back. He wanted the best men. The surface was snow, and the distances limitless. How far could he go? What new could he accomplish? Scott was thrilled with how well the sledge moved under the power of these men. It seemed to come alive under the horsepower of Petty Officer Edgar Evans and Stoker William Lashly. They would head off outward on to the Polar Plateau and Skelton's team would head back.

    The team moved ahead in a machine-like fashion, averaging 12 miles a day for the next week with almost no conversation and no stopping from meal to meal. All snow conditions and sastrugi varying in wavelength, amplitude, and direction were met head on, sometimes capsizing the sledge, leading to losses of stove fuel from the poorly sealed, corked containers. Temperatures varied throughout the day from –25°F during marching hours to –40°F at the lowest Sun angle at night, but the incessant drying wind cut into them, producing cracks in lips, cheeks, noses, and hands. Evans in particular seemed more vulnerable, with deep gashes on each side of his thumb as if made by a knife. Sudden falls and being jerked around by their harnesses in rough sastrugi shook them up, but good mileage was maintained. By the end of the month, Scott calculated from the food supply they would have to turn back. Now out some 200 miles on the Plateau, he had hopes that somewhere a surprise discovery would loom up from the vast whiteness. Perhaps another mountain range, perhaps a coastline, perhaps a deep hole in the ice, anything, anything at all to set him apart from Armitage and all the others in having seen something special with only his eyes. But by Monday, 30 November, there was nothing, only more whiteness. His poetic, naturalist, and gloomy side began to surface:

But, after all, it is not what we see that inspires awe, but the knowledge of what lies beyond our view. We see only a few miles of ruffled snow bounded by a vague, wave horizon, but we know that beyond that horizon are hundreds and even thousands of miles which can offer no change to the weary eye, while on the vast expanse that one's mind conceives one knows there is neither tree, shrub, nor any living thing, nor inanimate rock – nothing but this terrible limitless expanse of snow. It has been so for countless years, and it will be so for countless more. And we, little human insects, have started to crawl over this awful desert, and we are bent on crawling back again. Could anything be more terrible than this silent, wind-swept immensity when one thinks such thoughts? *(Scott, 1905, v. II, pp. 264–265)*

**Heading Back**  On Tuesday, 1 December, they began the return march and the immediate effect of now having the wind at their backs was a relief, but "... new difficulties soon appeared." Overcast skies set in, making progress slow and cumbersome and "... the sky eventually became so gloomy that we were forced to camp and sacrifice more than an hour of the afternoon." Wednesday started slightly better with occasional weak glimpses of the Sun, but soon deteriorated "... and the light became shockingly bad." Encountering high sastrugi, the sledge repeatedly capsized, and they fell so frequently that Scott feared one of them might be injured and camp was made in the middle of the day.

As Scott lay curled up in his bag he ran over things in his head, again and again, brooding. Fourteen days of food left and maybe twelve days of fuel, capsizing and poorly sealed containers had eaten fuel. Should food be cut back? Should marches be made longer? Scott didn't know their tolerances and limitations. They could easily have done both, and with the distances and time at hand they could even have gone without food for several days, if not a full week. Still, he fretted.

If we could get clear weather, I believe we have not overestimated our marching powers in supposing we can cover the longer daily distance required to reach the safety of the glacier, but this overcast weather puts an entirely new complexion on the matter; it is quite clear that we cannot afford delay. I don't like to think of half rations; we are all terribly hungry as it is, and I feel sure that we cannot cut down food without losing our strength. I try to think that at this altitude there cannot be long spells of overcast weather, but I cannot forget that if this condition should occur frequently we shall be in "Queer Street." *(Scott, 1905, v. II, p. 268)*

He had been away from the other, returning party for scarcely 10 days and had come from an emotional point of thrill and excitement to one of almost despair. After a couple hours of in the tent, Evan looked out to find the Sun shining. Leaping out, rapidly packing up, they were off in 10 minutes and now found the same surface to be not so bad after all. They sped right along. Evans and Lashly were steady.

Nearly 100 miles had been covered and at this rate in another six days or so Depot Nunatak would be raised on the eastern horizon. Fretting continued, even though the surface was comparatively good and good progress was being made. Yet on the 8th, a surface that Scott may have recalled from the outward journey, of mixed patches of snow and ice, was now covered by a recent snowfall hiding the ice, making walking unsure and unpredictable. The crampons were brought out, but the friction on the sledge runners was "terrible." It seems even a small victory could not be found, regardless of conditions. On Wednesday, the 9th, with sledge weight at a minimum, Scott felt "... everything going wrong for us, and the marches on that day and those that followed I can never forget." The sledge pulled like a barked log. Scott felt the fear, gloom, and danger coming down on him. He felt the lives of these honest, hardworking, kind, and trusting men as a heavy weight as bad as the sledge itself. He talked it all over and explained that they needed to conserve stove oil by having cold lunches and a fried breakfast.

The gravity of the situation preyed continually on Scott. Aside from hauling the sledge, all was on his shoulders. On board ship he had a cast of subordinates with which to consult with wide margins of error for daily navigation and meals. Deeply buried in his bag each night, wide-awake, he thought: They might well at any hour fall off the edge of the Earth. Images of the huge, formidable mountain wall of the Royal Society Range ran through his mind. Where would they come out? Would they be lucky enough to find Armitage's secret passageway?

Thursday, 10 December, began again with pulling so difficult that Scott found it "... impossible to drag one's thoughts away to brighter subjects." Trudging along toward lunchtime, saying little to one another, the prospect of a "... cold, comfortless, lunch ..." only heightened Scott's anxieties. Early in the afternoon, Evans, with his sharp eyes, sighted wisps of land on the horizon. The pace quickened and soon nunataks appeared on "... both bows ..." Scott was cheered, the surface felt better, the sledge moved easier, and "... we struggled on through the remainder of our march with renewed hope." More peaks rose to the southeast, but Scott's worries rekindled when he could not recognize any of them. On the way outward he had yet to adopt the habit of periodically stopping, looking back, and studying where he had been. On the ocean this was not needed but here it was critical.

I imagine we are too far to the south, but I am not at all certain. I rather thought that when we saw the land it would bring immediate relief to all anxiety, but somehow it hasn't. I know that we must be approaching the edge of the plateau, but now the question is, where? *(Scott, 1905, v. II, p. 275)*

The next two days did little to help matters. Overcast skies and low clouds only afforded furtive glimpses of the peaks here and there. Daily mileage even with longer working hours went down to 8–10 miles, but land didn't seem to get much closer. And even though camp on the 12th was near where Skelton had turned back, although 15 miles farther south, Scott was "... still left in horrible uncertainly as to our whereabouts, as I could not recognize a single point." And worst of all, Scott felt they had been giving their most and were being reduced to "... gaunt shadows of our former selves. My companions' cheeks are quite sunken and hollow, and with their stubby untrimmed beards and numerous frost-bite remains they have the wildest appearances, yet we are all fit, and there has not been a sign of sickness . . ." (Scott, 1905, v. II, p. 276).

He was thinking of scurvy. But it had only been a little over two weeks since, fit as fiddles, they had left Skelton and the others, and here they were evidently mere ghosts of their former selves. How about going on half rations for six weeks as Shackleton and his men would do four years from now? This was another dimension of Scott's fear: that they were drying up and wasting away right in front of themselves. When they did return to McMurdo, they had collectively lost 40 lbs in weight, which is perfectly normal and expected for men of this size who had worked this hard at these elevations for 59 days. He especially lamented the fact that he and Evans had run out of pipe tobacco, even though they had carefully put themselves on half rations. But the fault was not theirs: ". . . it was our long stay in the blizzard camp that reduced us to this strait."

Scott was by now at the final phase of fear and desperation. He just wanted it to end whatever the outcome. "There is one blessing; the next day or two will show what is going to happen one way or the other. If we walk far enough in this direction we must come to the edge of the plateau somewhere, and anything seems better than this heavy and anxious collar work" (Scott, 1905, v. II, p. 276).

Sunday morning, 13 December, came with a strong southerly wind with ". . . blinding drift." They had to press on regardless of the weather and cover respectable ground, but after four hours of steady marching Evans's nose froze up yet again, but they had to keep going.

By morning, Monday the 14th, the storm had passed and there was much land in sight, none of which unfortunately Scott could identify. "In this bewildered condition we packed our sledge and continued on eastward." What else could they do? In a short time, the ice began showing signs of major disturbances ahead and before long they felt the surface begin to descend. It would be a tumultuous, harrowing, and exhilarating day.

**Eureka on a Thread**   With the ice becoming heavily crevassed and hummocky, they proceeded cautiously not knowing where to go. Skirting back and forth, no place looked good. The wind and spindrift increased. They huddled together and talked it over with no result except the collective continuing confusion of not knowing where they might be. Although Scott realized the danger in blindly heading into the twisted ice without a clear idea of what was coming, he was also worried that if they made camp here and waited for things to clear they might get stuck in another "Blizzard Camp" and face certain starvation. He was anxious to resolve things one way or another. He had had enough. He presented the case to Lashly and Evans. Onward, straight at it, was their verdict, just like Lord Nelson.

The icy slope quickly grew steeper, and Scott put the men out behind to hold back the sledge as he led on. It wasn't long before Lashly suddenly slipped onto his back and began a freefall down slope. In an instant Evans was jerked off his feet and was cascading along, tumbling after Lashly, the sledge fully out of control. The whole outfit hurled past Scott ripping him off his feet and dragging him along willy-nilly as a confused mass of men, ropes, and sledge. The glassy surface suddenly steepened into a shelf, briefly launching them all only to suddenly land on rougher ice and continue rolling and tumbling. Their velocity steadily increasing until in an instant they landed, sliding on a more gradual snow-clad slope where bumps and irregularities caught them and brought the whole mass to a slow halt. Scott struggled to his feet, felt in one piece, and looked about, certain that in their haste he had committed the worst crime of exploration: bringing serious harm to the whole enterprise through faulty leadership. He was certain the others must have broken a limb or two. What would he do now? "Then to my joy I saw the others also struggling to their legs, and in another moment I could thank heaven that no limbs were broken" (Scott, 1905, v. II, p. 280).

Oh! What a relief he felt. Each was heavily bruised on the legs and arms and Lashly had caught it bad on his back with heavy bruising. But they were intact. Heavy clothes had protected them. The best was yet to come. For as Scott regained his senses and began looking around, he saw they had fallen some 300 feet. The spindrift was blowing around them at this lower level and out to the east the sky was clear with limitless visibility.

As soon as I could pull myself together I looked round, and now to my astonishment I saw that we were well on towards the entrance of our own glacier; ahead and on the either side of us appeared well-remembered landmarks, whilst behind, in the rough broken ice-wall over which we had fallen, I now recognised at once the most elevated ice cascade of our valley ... all around us the sky was bright and clear and our eyes could roam from one familiar object to another until far away to the eastward they rested on the smoke-capped summit of Mt. Erebus. *(Scott, 1905, v. II, p. 280)*

The veil had been lifted. They had miraculously, chaotically tumbled into the promised land. And there was good old Mt. Erebus signaling the way home, a sentinel to safety. They would survive after all. Scott felt deeply redeemed and overjoyed.

I cannot but think that this sudden revelation of our position was very wonderful. Half an hour before we had been lost; I could not have told whether we were making for our own glacier or for any other, or whether we were ten or fifty miles from our depot; it was more than a month since we had seen any known landmark. *(Scott, 1905, v. II, p. 281)*

And now good old Depot Nunatak was just down the valley welcoming them "... where peace and plenty awaited us." But the day's excitement was far from over. The full price for admission to the Dry Valleys had not yet been paid. First there was some critical work to be done. The sledge had mysteriously not capsized until the end, but items had fallen from it all along and the food box had opened scattering what was left here and there. Everything was gathered, although much of the food scraps were lost, the sledge repacked, and they set off again all out front for Depot Nunatak some four or five miles away. Evans pulled on the left and Lashly on the right of Scott in his usual center lead position. Ahead was a fairly long shelf leading to another icefall below which was the shelf containing Depot Nunatak. They had not gone far when they emerged from the wind protection of the icefall; the southerly wind hitting broadside on the sledge made it slide sideways on the smooth ice. Soon after Scott asked Lashly to fan out up wind to counter the wind, Scott and Evans suddenly dropped from sight into a vast lightly bridged crevasse. Lashly immediately hit the deck and dug in his heels. The sledge streaked ahead out over the crevasse snapping off the front end of the left runner under the weight of Scott and Evans hanging below. It all happened so fast that neither Scott nor Evans had any sense of what happened. The first thing he knew Scott found himself hanging some 12 feet below in darkness. Taking off his snow goggles, he looked up to see Evans hanging just above him. He wiggled around and felt his foot strike a shelf on which he gained some support. The shelf was a local wedge-like slab of ice forming a small oasis in an otherwise bottomless crevasse. Scott felt again redeemed. He directed Evans to slip down in his harness, letting it ride up under his arms, and also gain a footing on the shelf.

Up on the surface Lashly, not knowing what had happened below, held on for dear life and, keeping full tension on his hauling rope, edged himself over to the sledge, slid out a pair of skis, and placed them as a bridge under the sledge and across the crevasse. This stabilized things a bit, but the arrangement was precarious; each time he relaxed his grip the sledge moved ahead. He was not going to be able to toss down a rope and pull anyone up. After a few words back and forth it was clear that either Scott or Evans would have to swarm or climb the harness rope back to the surface. This was not going to be easy. Even without the added weight of heavy clothing, in warm, barehanded temperatures, this was not easy. Nevertheless, Scott stepped up to the challenge. He was 34 and, although hardened from all the hauling, was by no means a regular at doing anything like this since his cadet days. He unbuckled his harness, removed his mitts, swung them over his shoulders, gripped the rope and headed up. After 4 feet he could get his feet into the belt of his hauling harness, giving him a breather, but he couldn't tarry with bare hands. Presently he began again and after another 4 feet he reached the rope loop or stirrup from the sledge tongue to which the hauling lines were attached. He got his feet into this loop, rested a bit, and then raised himself again until he caught hold of the sledge itself. Pulling and kicking with his crampons into the wall he levered himself up and out and onto the surface. Lashly said, simply: "Thank God!" It was then that Scott saw how dangerous a predicament they had been in. Thrusting his hands into his jacket, under his armpits for a short while and then donning his mittens again, he unhitched his harness and used it as a belay rope to Evans. With Evans's weight off the sledge, he and Lashly pulled Evans to the surface.

In their euphoria at finding the way home they had been reckless and had nearly paid the full price for it. Now as they marched ahead, past Blizzard Camp, eyes were pealed and hauling positions were staggered to guard as much as possible from losing more than one man down a crevasse. They were learning. By 6 o'clock the towering brown-black columnar cliffs of Depot Nunatak were reached, and camp made in brilliant, above 0°F, wind-free weather. It was a relief to cast eyes on anything but shades of white. Scott was elated and felt finally that he had perhaps paid his dues; his fight against Nature was over. From now on he would listen and watch and be careful, maybe, or at least for a while.

Supplies left here were meager, but there was more than enough to get to the big depot off Solitary Rocks, and there was also a nearly new 9-foot sledge, left by Armitage the previous year, which they were pleased to get. In the fine sunny weather and the huge emotional relief, the bruised, stiff, and worn men "dawdled" over every motion around camp. Lashly's over the stove musical ditty now seemed to make sense, Evans waxed on reflectively on the miracles of surviving snow bridges, and Scott himself reflected "... that in the brief space of half an hour we passed from abject discomfort to rest and peace." Such times, in seemingly coming from a cauldron of certain hell into a sweet green meadow, when the saving grace is due far more to naive chance than any obvious skill, is a sensation defining, to some, life itself, and Scott clearly felt this. It can also kindle and breed bravado. And as the sledge pulled away from Depot Nunatak on Wednesday afternoon there was a new swagger in their walk. "This sort of work was mere child's play to our hardened muscles, and that night we reached the broad amphitheatre below Finger Mountain" (Scott, 1905, v. II, p. 287).

The big depot opposite Solitary Rocks was reached the next day, where ample food was found, and here as at Nunatak Camp, Scott found messages from Ferrar and Skelton telling of the safe arrival of these parties, which sustained his emotional high spirits. Continuing now on old familiar ground, camp on the 16th was made among the blocks and rubble near Armitage's Knob Head Moraine campsite. Scott now had a decision to make. He could hustle back to the ship and make sure every effort was being made to dislodge *Discovery* for the voyage home, or he could bask in the afterglow of rebirth and take some time to explore and examine the surroundings. The common desire is to lie about on the beach, still in the wilderness, and sustain what may be some of the purest sensations felt by man. The pervasive emotional warmth of dwelling near, but securely beyond the mouth of the lion is too much to fight. Scott could look back anytime he cared, up over his shoulder to the Plateau, see the ground blizzards whirling up, and feel the scars somewhere within him rekindling the most desperate fears and anxieties.

**Discovery of the Major Dry Valley**    Although it seemed much longer, they had been away from Winter Quarters for only 50 days; a relatively short time. This was it, the end of the expedition, no telling what Scott would be stuck with back at home, Christmas was hardly a week away, and the ship could be reached in two or three days from here. No need to hurry back. Let's go exploring, thought Scott. There was only one obvious place to go in this basin. They had come in here along the "South Arm" of the glacier, Ferrar's Glacier, but there also seemed to be a "North Arm" to the glacier that went across in front of Solitary Rocks and then possibly curved eastward back toward McMurdo Sound. The two arms being separated by a long sharp ridge resembling to Scott's officers a Gurkha knife, hence named Kukri Hills. It might be fun to look around here, on the far side of the Kukri Hills. This they did. There had been discussions ever since Armitage found this basin on where all the ice went. Ice flowed in here from many locations, the most massive of which was down from the Polar Plateau itself. But where did it all go? Where did it exit? Some went down in Ferrar's Glacier, but surely not enough to balance all the inflows. Maybe the North Arm was the answer. Scott was curious. His naturalist explorer side came alive.

So, on Thursday morning, 17 December, the party began to descend to the north following the ice. By the end of the day, after 8 or 10 miles, the ice, indeed, turned strongly to the east, straight for McMurdo, and became increasing rough to the extent that the sledge was in enough danger to make camp and do further exploring simply on foot. What was more perplexing, however, was to see in the distance the glacier apparently diminishing in volume and in the far distance beyond it ". . . stood a lofty groin of rock which seemed a direct bar to its further passage." What a surprise. What could that mean? Much to Scott's credit, his bafflement was exceedingly well founded. Scott was finally going to have his chance to discover something of timeless importance and beauty.

By 7 o'clock the next morning, Friday the 18th, they were off down the glacier sans sledge, shod with crampons, carrying an alpine rope, and pockets filled with chocolate, sugar, and biscuits. Rucksacks would have made the day. Moving eastward the glacier began to fully break down into a series of fingers between which ran channels of melt-water in icy beds. Getting down in one channel they marched along and found the ice dramatically diminishing until it ". . . gradually

Figure 2.21 The end of the Taylor Glacier in Taylor Dry Valley, first discovered by Scott, Evans, and Lashly on 18 December 1903

dwindled to this insignificant termination." Beyond was a shallow frozen lake into which the melt water poured, moving under the ice (Figure 2.21). This was, rightly so, astonishing:

So here was the limit of the great ice-river which we had followed down from the vast basin of the interior; instead of pouring huge icebergs into the sea, it was slowly dwindling away in this steep-sided valley. In fact, it was nothing but the remains of what had once been a mighty in-flow from the inland. *(Scott, 1905, v. II, p. 290)*

They had to go on. Climbing down to the base of the dying glacier they found a layer of fine mud 10 or 12 feet thick, which Lashly immediately remarked: "What a splendid place for growing spuds!" Mysteriously enough, time would reveal that with nothing plant-like whatsoever alive in this region ever, the mud contains virtually nothing organic from which to build nutrients, and spuds would not sprout. Skirting the rough ice on the lake[9] (West Lake Bonney), they came to the

---

[9] This was the start of the discovery of many "frozen" lakes of the Dry Valleys region. They are mainly saline brines (most often calcium chloride brines) with low freezing points and high pH (~12) values. Some reach depths of over 30 meters. With lighter freshwater on top, they are typically frozen over and act as solar collectors absorbing summer sunshine and becoming, in the base level of the deeper lakes, as warm as ~80°F. Algal mats of beautiful cyano-bacteria populate the floors of many lakes and, overall, the lakes are a constant focus of research (e.g., Faure & Mensing, 2010, p. 728, etc.). Lake Bonney was named by geologists of Scott's next,

Figure 2.22 Lake Bonney groin in Taylor Dry Valley

transverse ridge or groin (actually a riegel) thrusting out from the foothills of the Kukri Hills on the right and found a narrow channel cutting through to the northwest (Figure 2.22). Only 17 feet wide at its narrowest, with high cliffs on each side, the valley below immediately opened up again where the floor was filled by another massive frozen lake (East Lake Bonney) "... a mile in breadth and three or four miles in length." Skirting again along the shore, Scott was fascinated to be moving into an entirely ice-free region of sand and rock, when only 50 miles west stood a true harsh polar wilderness. At the end of the lake the valley floor rose on "... a mass of morainic material" deposited in long ridges, eventually becoming re-deposited under the action of numerous small streams coming in from the valley walls. And within a mile or so the moraine materials ceased altogether to be replaced by fans of alluvial sands from small active streams and rivulets supplying puddles and ponds fully free of ice. Scott, in his enthusiasm, had been loping along in this Eden when Evans calmly asked if they needed to keep carrying the lunch. They sat together on a

*Terra Nova*, expedition (1910–1913) for Professor Thomas Bonney, Geologist at University College, London. And this valley itself, Taylor Valley, and the glacier leading into it all the way down from the Polar Plateau was later named after Griffiths Taylor, geologist with Scott's *Terra Nova* Expedition.

... small hillock of sand with a merry little stream gurgling over the pebbles at our feet. It was a very cheery meal, and the most extraordinary we have had. We commanded an extensive view both up and down the valleys, and yet except about the rugged mountain summits, there is not a vestige of ice or snow to be seen; and as we ran the comparatively warm sand through our fingers and quenched our thirst at the stream, it seemed almost impossible that we could be within a hundred miles of the terrible conditions we had experienced on the summit. *(Scott, 1905, v. II, p. 291–292)*

Scott was eager to see more and possibly get a glimpse of the Sound in the hope that the ice had left, giving *Discovery* a fighting chance of escape. Down the valley they went, into the deepest part of the gorge where a heavy boulder field blocked the way, beyond which another groin (Nussbaum Riegel) extended out from the Kukri foothills blocking any view to the Sound. After a long, difficult climb of 700 feet around and over sharp boulders the top was finally reached ". . . but, alas! Not to catch any glimpse of the sea"; The valley continued on winding its way back and forth with more frozen lakes in a northeasterly direction for as far as the eye could see, ". . . and some five or six miles below yet another groin shut out our further view."

But from our elevated position we could get an excellent view of this extraordinary valley, and a wilder or in some respects more beautiful scene it would have been difficult to imagine. Below lay the sandy stretches and confused boulder heaps of the valley floor, with here and there the gleaming white surface of a frozen lake and elsewhere the silver threads of the running water; far above us towered the weather-worn, snow splashed mountain peaks, between which in places fell in graceful curves the folds of some hanging glacier. *(Scott, 1905, v. II, pp. 292–293)*

Scott had more than simply a poetic eye and a good hand with words; he was a keen observer of landscape. He saw delicate key and critical features in the terrain telling about past times when things were different. He was remarkably not caught up in a world of fixed-ness (Figure 2.23). "The colour was therefore predominant, but everywhere at a height of 3,000 feet above the valley it ended in a hard line illustrating in the most beautiful manner the maximum extent to which the ice had once spread" (Scott, 1905, v. II, p. 293).

It is not a simple matter to see a lifeless terrain for what it truly is at the moment it is first observed, especially by those not cautioned to be on the lookout for something specifically of this ilk. And, other than the special feature of being bare ground, there was little here for the ordinary man to grasp to convince oneself that here was something truly special. But Scott did see into another dimension.

I cannot but think that this valley is a very wonderful place. We have seen to-day all the indications of colossal ice action and considerable water action, and yet neither of these agents is now at work. It is worthy of record, too, that we have seen no living

Figure 2.23 Taylor Valley, looking east toward McMurdo Sound, with Lake Bonney in the foreground and the Kukri Hills on the right

thing, not even a moss or a lichen; all that we did find, far inland amongst the moraine heaps, was the skeleton of a Weddell seal, and how that came there is beyond guessing. It is certainly a valley of the dead; even the great glacier which once pushed through it has withered away. *(Scott, 1905, v. II, p. 293)*

It was now getting late in the afternoon, nearly 4 o'clock, when Scott turned the party back west toward camp on the ice. They had seen a lot, discovered, and appreciated in real time what would become the single most studied and celebrated valley in all of Antarctica. Evans and Lashly probably wondered what all the fuss was about. The state of the ice in the Sound was not learned, this would have to wait for a few more days. It was a long hike back to camp, but they were in good shape, arriving near 10 P.M., having been on foot for 14 hours and covering over 25 miles. The next morning, Saturday, they reversed the trek first heading back west and south up the glacier and then around the nose of the Kukri Hills and on down to the depot beneath Cathedral Rocks, which they reached on the evening of Sunday, 20 December.

Looking out from camp, the Sound was all white, no deep field of black or blue telling of open water. Gaining the next morning the mouth of New Harbor, only in the far distance over 10 miles from Hut Point could a thin strip of blue be discerned.

Another surprise awaited them at Butter Point on the south side of New Harbor when no seals for supper could be found. The habit had already started of leaving a good supply of butter here on the way west to be used homebound to marinate fresh seal meat. All had greatly looked forward to this for the past eight weeks. No such luck today, but good old Skelton had anticipated this need and concealed "a buried treasure in the shape of some tit-bits of an animal which they had killed."

With Christmas day coming up on Friday, there was still a little time to spare before arriving with the best possible splash on Christmas Eve, and Scott wanted to see the curious moraines they called the "Eskers" just to the south that Armitage had found the previous season. Scott had a special interest in glacial features and a knack for seeing them for what they were. They went there on Tuesday, the 22nd, and spent half a day rambling about the craggy, bouldery ice looking for skua eggs for breakfast. Eggs were scarce, none could be found, but Scott enjoyed exploring.

Taking their time getting going the next day, on Wednesday afternoon Scott, Lashly, and Evans, still intact, still friends, headed off for the last time across McMurdo Sound and camped for the night a good day's march from *Discovery*. The weather was calm, and Scott must have enjoyed, musing in his thoughts over supper as Lashly cooked, the knowledge of having stepped far out onto the Plateau, much farther than anyone had been, survived, and found, to top it all off, a singularly special valley. And, so, late on Christmas Eve they saw the masts of *Discovery* in the distance and arrived to find, perhaps disappointingly, only four men at home: Koettlitz, Handsley, Quartley, and the "ship's steward" (Ford). All the others were out on the ice-sawing work detail; Armitage had all well under control. Freeing *Discovery* in time to head north before winter did not look good, but there was still time.

The ice edge was about 20 miles away and Armitage had a crew sawing in 10-foot-thick ice, but things looked hopeless, even when using explosives. In early January two relief ships, *Morning* and *Terra Nova* appeared on the ice edge. Things looked bleak for freeing *Discovery* and the directives carried by the relief ships where "to abandon the 'Discovery' if she could not be freed in time to accompany the relief ships to the North." On 15 January they began in earnest transferring the collections and gear to the relief ships. On the 18th easterly winds came, creating a roll that broke off several miles of ice; hopes rose, but there was still another 13 miles to go. This saddened Scott, who despaired that all was going poorly for them. But on the 28th *Discovery* began to creak and move a little; it was a false alarm. The agreement was made to leave by 25 February and, on the 10th, Scott gave the order to abandon ship:

I have made every arrangement for abandoning the ship. I have allotted the officers and men to the relief ships and drawn up instructions for the latter. The "Morning," I think, ought to be outside the strait by the 25th, but the "Terra Nova" with her greater power can remain perhaps a week longer. I don't think I ever had a more depressing evening's work. *(Scott, 1905, v. II, p. 345)*

Then suddenly on Sunday 14th a southeast wind started breaking up the remaining ice, and by 10:30 P.M. both relief ships were within a half mile of *Discovery* and ramming by *Terra Nova* soon brought her alongside *Discovery*. There was still much to do to free *Discovery* from the remaining ice collar about her and then get her off the grounding sediments she had become sandwiched into by the currents over two years. But even in the face of nearly wrecking *Discovery*, next morning, Thursday 18th, with the wind swinging around to the southeast and the Sound becoming tamer, *Discovery* headed north, Scott looked over at Hut Point and the little bay that gave so much protection and yet in the end nearly took everything. The ships rendezvoused on the lee of the ice tongue, where provisions, clothing, and coal was passed to *Discovery*. Everyone, officers and men, worked hard hauling sacks of coal and boxes of provisions from 2 P.M. until midnight, bringing on board 17 tons of supplies and 60–70 tons of coal; a relatively scant amount of coal. There would be more troubles, but Scott certainly returned triumphant. He didn't get to the South Pole, but they had founded a ground for an everlasting research colony and had discovered a great deal, especially the Dry Valleys.

### 2.2.2.3 Scientific Results

In terms of raw manpower, the expedition was grossly understaffed in scientists. The science covered geomagnetism (Louis Bernacchi), marine biology (Thomas Hodgson), botany (Reginald Koettlitz), zoology (Edward Wilson), and geology (Hartley Ferrar). There was not much botany to be done, but Koettlitz, chief surgeon, doubled also as a geologist of sorts. Bernacchi, Hodgson, Wilson, and Ferrar did the bulk of the science, which overall was certainly respectable. The key aspect of the results was in sharing specimens and measurements with experts back in England, which led to sound, valuable, and comprehensive reports published in large folios beginning in 1907. And, perhaps above all, the terrain had been explored in some detail, especially the inroads made by Armitage, Scott, and Ferrar in opening routes into the Dry Valleys (Figure 2.24).

Almost everyone became sensitized to noticing geologic features, and many seminal observations were made regarding past ice coverage and rock formations wherever they went. Scott himself had a good and curious eye for the general aspects of Nature and relished being able to visit locations where close observations could be made, as is exemplified by his map of visited locations and the general geology of the whole area (Figure 2.25). This certainly furnished a valuable overview of what might be investigated in future work. But it is the geologic map of Hartley Ferrar that showed what the Dry Valleys have become noted for (Figure 2.26). A thorough petrographic and chemical analysis was made of Ferrar's rocks by Prior (1907) of the British Museum, who also examined the rocks from Ross's expedition to this region (Prior, 1899) and the rocks of the Southern Cross expedition (Prior,

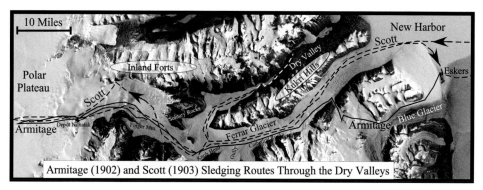

Figure 2.24 Sledging routes taken by Armitage in 1902 and Scott in 1903 in reaching the Polar Plateau and discovering and exploring the Dry Valley

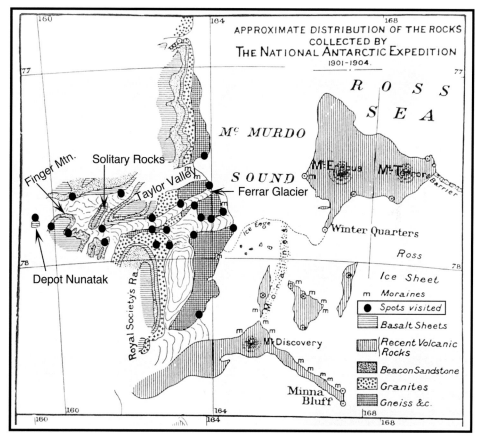

Figure 2.25 Preliminary geologic map showing principal features and specific locations (black dots) visited by Scott and Ferrar and where samples were taken by Ferrar (after H. T. Ferrar, in Scott, 1905, v. II, following p. 448)

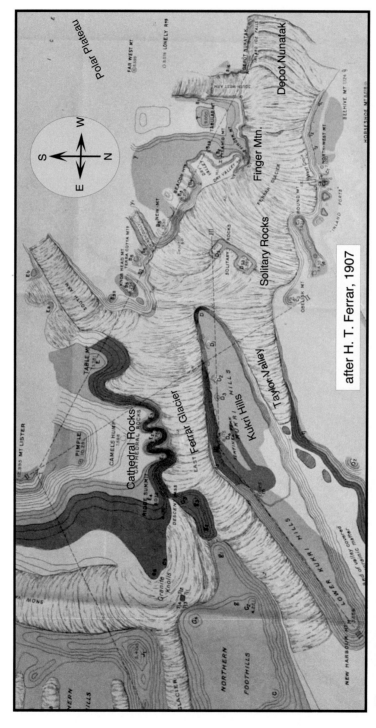

Figure 2.26 Geologic map of the Dry Valleys region by H. T. Ferrar (1907)

1902). His analysis of the Knob Head rock shows a lightly elevated MgO content, which is consistent with the later close petrography of the pyroxenes by Benson (1916), described in Section 2.2.3.1. Ferrar's dolerite samples are not particularly revealing because they were mostly taken very near the contact with the Beacon and are strongly chilled and fine grained. In the Kukri Hills area, Ferrar found a location where dolerite is against granite and the granite shows wisps or small tongues penetrating the dolerite. Although in general he saw that the dolerite was later than the granite, here he concluded that this granite was younger than the dolerite (Ferrar, 1907, p. 36). Little did he, or anyone else at this time, realize that the dolerite was hot enough to remelt and locally mobilize the granite; as will be abundantly seen in Chapter 6, this is abundant throughout this region.

Hartley Ferrar's geological work was certainly, in the end, more than adequate, even admirable, especially in light of having no topographic maps or aerial photographs to use and having to cover all aspects of geology and geomorphology from ice to volcanism. In this geologically magical region where everything is so wonderfully exposed, it might have been hard to fail no matter who did the initial work. Much as landing on the Moon, any handful of rocks, photographs, and simple description of the terrain was scientifically epic making. Originally the Royal Society had wanted a strong hand in the whole expedition, which would have been designed around doing high caliber science, but this would have relinquished some control. First class seasoned geologists, presumably, would have done better, but maybe not, they may have become mired in the details. Ferrar (28 January 1879–April 1932) had just recently received his BS degree in Geology from Cambridge and was the youngest member of the scientific staff. He was cutting his teeth here, much as Darwin did on the *Beagle*. His fresh enthusiasm is clear in his writing, which covers a great deal more than the Dry Valleys, covering everything geologic from the Hut Point–Erebus volcanics to the landforms and glaciology. Oddly enough, no one climbed Mt. Erebus or went to find the South Magnetic Pole; Shackleton's geologists would do this.

### 2.2.3 Ernest Shackleton's Nimrod Expedition (1907–1909)

The bitterness Shackleton felt in being invalided home manifested itself in his own National Antarctic *Nimrod* expedition of 1907–1909. A merchant mariner, used to making trips between England and South Africa, and an expert in organizing and packing cargo, Shackleton put together a small, lean, and tough expedition. Fifteen men: seven scientists, a cartographer, two surgeons, a mechanic, a cook, a couple of general helpers, and a leader were to spend one year doing science and making an attempt on reaching the Pole. Each man was chosen to be rough, ready, and eager. Besides Shackleton himself, Frank Wild and Ernest Joyce had been here with Scott as Naval Seaman, and they had seen and done a lot.

Unable to find any other suitable place to establish a base, Shackleton had no choice, breaking his promise to Scott, he headed back to McMurdo Sound, which was choked with ice. There was no chance of getting back to Hut Point. Try as they might, waiting in vain for things to open up, camp was finally set up, building a hut, at Cape Royds 25 miles north of Hut Point; inconvenient, for in getting to Hut Point the formidable Ice Tongue had to be traversed if the Sound was not viable, but at least they were on solid ground. And Scott's hut would certainly be a vital staging location for all trips Southerly or into the Dry Valleys. Unloading began in early February, and *Nimrod* was off North by the 23rd.

The main intentional geological work involved ascending Mt. Erebus, trekking to the South Magnetic Pole, and an excursion into the Dry Valleys. But the most seminal and dramatic was Shackleton's traverse of the Transantarctic Mountains via the Beardmore Glacier highway, which he discovered during his attempt on the Pole, reaching the Polar Plateau and getting within about 100 miles of the Pole itself. Although none of his party were geologists, they each had an interest in geology: Eric Marshall, surgeon and cartographer, Jameson Adams, meteorologist, and Frank Wild, who had been with Skelton and Scott in the Dry Valleys all the way out to and beyond Depot Nunatak.

The ascent of Mt. Erebus took place on 5 March shortly after arrival. The summit party of T. W. E. David, Geology Chair at University of Sydney, Douglas Mawson, David's past student, and Jameson Adams, meteorologist, reached the summit on the 9th. The crater is unusual in containing an active lava lake with large crystals of euhedral anorthoclase floating on the surface, which have also been deposited on the ground by local eruptions (more later). They determined the summit elevation to be 13,500 ft., which is about 1,000 ft. too high.

The Western Party of Raymond Priestley, geologist, Bertram Armytage, general helper, and Philip Brocklehurst, assistant geologist, went to the Dry Valleys via the Ferrar Glacier, in hopes of finding fossils in the Beacon, which they found remarkably "unfossiliferous"; yellow, green, and black lichens were, however, found on the slopes of Knob Head. They hoped to ascend to Depot Nunatak, but fresh snow on slippery ice held them back and instead made a circumnavigation of the Kukri Hills, finding fault with Scott's and Ferrar's understanding of the Solitary Rocks:

An examination of the Solitary Rocks proved that the map was incorrect at this point. The previous expedition thought that the rocks formed an island, with the glacier flowing down on either side, but a close examination showed that the rocks were in reality a peninsula, joined to the main north wall by an isthmus of granite at least one thousand feet high. *(Shackleton, 1907, p. 64)*

True enough but walking on the ice over and back to Taylor Valley as Scott did, the connection to the Asgard Range via Pearse Valley is not all that obvious. Even with this realization their revised geologic map shows only a tenuous connection (Shackleton, 1907).

The Northern Party, consisting of David, Mawson, and Alistair Mackay, assistant surgeon, was charged with going Northwest to locate the South Magnetic Pole. Leaving in early October 1908, their route went first to New Harbor, at the foot of Ferrar Glacier, and then carefully along the coast for almost 300 miles, traversing Drygalski Glacier on the southern edge of Terra Nova Bay. They then headed inland northwesterly, sledging by man hauling for 350 miles, finding the Pole on 16 January at 72° 15'S, 155° 16'E at an elevation of 7,200 ft. The return journey took two weeks, arriving to find the Sound free of ice and miraculously the *Nimrod* showed up looking for them and, docking at the ice edge, "rescued" them [David gives an extensive report in Shackleton (1907, v II, pp. 73–222)]. *Nimrod* then went on to do the same for the Dry Valleys Party at New Harbor. A few days later, after a good deal of trauma, Shackleton's Pole party was similarly plucked off the ice edge at Hut Point on 4 March 1909 and the expedition headed north, all intact (Shackleton, 1907; Riffenburgh, 2004). Much to his joy, and possibly the chagrin of Scott, Shackleton was Knighted for his extraordinary accomplishment.

### 2.2.3.1 Scientific Results

Trudging up the Beardmore, Wild was immediately struck by the overall similarity of the formations he had seen in the Dry Valleys. He recognized the intimate prevalence of the dolerite sills in the Beacon Sandstone, all floored by granite in the bordering mountain walls. In the evenings after supper, they went to the moraines and walls, looked around, and collected samples. The first definitive fossils on this side of Antarctica (Cambrian *archaeocyaths*) were found by Wild in limestone erratics on the medial moraine here (David & Priestley, 1914). Ferrar had found some apparent fossils in the Beacon of the Inland Forts and near the Kukri Hills, but they were too well cooked by nearby dolerite to identify (see note by E. A. Newell Arber in Ferrar, 1907, p. 48). Shackleton and Wild made enough observations on the Beardmore to make a stratigraphic section covering some 1,500 ft., containing a number of thick coal seams, one sample of which contained fossilized coniferous wood (Shackleton, 1907, p. 299).

Besides the general overviews of the geological results given by David and Priestley on Antarctica itself (Shackleton, 1907, v. II, pp. 268–307) and Priestley on the Dry Valleys (pp. 315–321), there are brief reports on many subjects

including Erebus eruptions, ice physics and chemistry, meteorology, geomagnetism and the Aurora Borealis, and one page on rock mineralogy and chemistry by
Mawson. A much more extensive and thorough collection of geological reports are
included in volume II of the Reports on the Scientific Investigations of the
expedition that came out in 1916. There are reports on the Erebus volcanics and
xenolithic inclusions, the erratics at Cape Royds, pyroxene granulites, and massive
limestones; some of these later samples were collected in the medial moraine of the
Beardmore Glacier at 85°S by the Pole-seeking party.

   Most notable in the present context are detailed reports on petrology by Mawson
(1916) on all the rocks, including a brief section on the dolerites (p. 228), and a
brief (eight pages) but excellent petrographic report by Mawson's colleague, W. N.
Benson, at Sydney on the dolerites found as erratics in the Dry Valleys and on
Ross Island. Two mentions are notable: First he mentions finding dolerite as
xenolithic inclusions in Erebus lavas, suggesting dolerite might underlie Ross
Island; and second, he accurately describes large enstatite–augite intergrowths and
twinning in the Knob Head sample (#611), which is diagnostic of the existence of
the characteristic orthopyroxene concentrations in the Basement Sill and has
slightly elevated MgO at 8.64 wt.%, as found by Prior (1907). [This sample seems
to have had a life of its own as it was examined or described by others (e.g.,
Browne, 1923).] The rocks collected by the Northern Party in their excursion along
the west side of McMurdo Sound on the way to the magnetic pole were cached at
Depot Island, just north of Granite Harbor, and were not accessible at the close of
the 1909 season. These rocks were later retrieved by R. F. Scott's Northern Party
in 1912, as described presently, and were examined and briefly reported on by
Cotton (1916); these are some older basaltic dike rocks and foliated granite from
Depot Island. [In this overall context of petrology and mineralogy, the brief report
by Stewart (1951) on a survey of the "Mineralogy of Antarctica" is particularly
useful and interesting, including a comprehensive bibliography.]

   Additional comprehensive expedition Reports on the Scientific Investigations
on Biology and Geology (v. I) came out in 1914. The Geology volume by David
and Priestley (1914) is massive (over 300 pages), deeply comprehensive, and has a
wealth of information on the evolution of the Antarctic ice cap, glacier dynamics,
ice growth and erosion, physiography, volcanism, and possible tectonic evolution
of Antarctica in its relation to the other continents. It is laced with excellent
insights on many fronts, including the origin of the Royal Society Mountains as a
large Horst of indeterminate, but perhaps young, age and correctly laying out the
age sequence from the Pre-Cambrian basement rocks, gneisses, and granites,
through to the top of the Beacon Sandstone Devonian to Cretaceous. A suggested
cross-section from the magnetic pole to the coast is particularly insightful for its
time (Figure 2.27), although the dolerite sills are left out. And the general

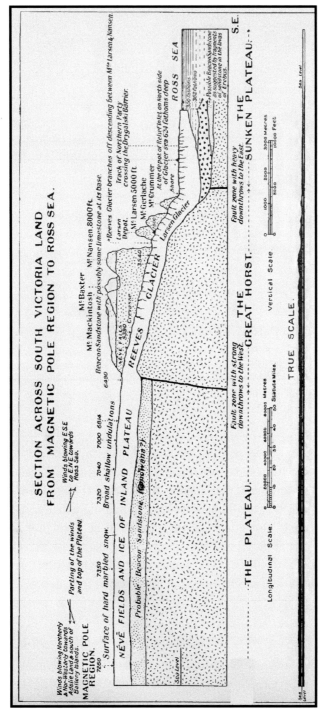

Figure 2.27 A suggested cross-section from the magnetic pole to the coast by David and Priestley (1914), which is particularly insightful for its time, although the dolerite sills are left out

67

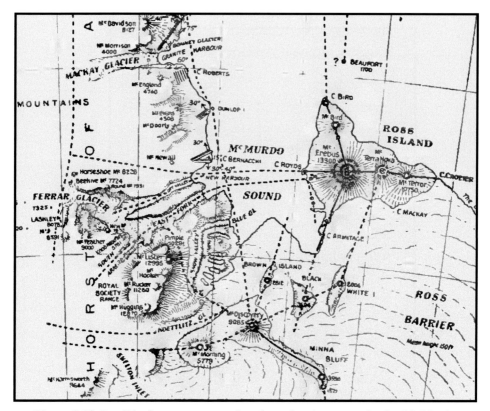

Figure 2.28 Possible fractures connecting the volcanic centers in the McMurdo region, as suggested by David and Priestley (1914)

McMurdo physiography and possible tectonic or fault-related relations between the various volcanic centers of Mt. Erebus, Black Island, White Island, Mt. Discovery, and Mt. Morning is presented in some detail (Figure 2.28), along with extensive remarks on the growth, evolution, and ongoing activity of Mt. Erebus. The many photographs and drawings are interesting and comprehensive. Their general, glaringly incomplete, understanding of the nature of the Dry Valleys, which is not restricted to this group, persisted until almost 1960, is also rather revealing, and will be further mentioned in the grand summation of this work.

### 2.2.4 *R. F. Scott's* **Terra Nova** *Expedition (1910–1913)*

Scott's ill-fated *Terra Nova* Expedition of 1910–1912, with geologists Frank Debenhan, Griffith Taylor, Raymond Priestley, and Charles Wright, who was a glaciologist, carried out a large program of mapping, fossil collecting, and examining Mt. Erebus. It is on this expedition that Captain Scott and his four companions (Oates, Evans, Bowers, and Wilson) perished on their return to McMurdo after reaching the South Pole in January of 1912. They, nevertheless, collected 35 lbs. of valuable samples from the Beardmore moraine below Mt. Buckley, which were carried to the end. This moraine, rich in rocks from the Beacon Sandstone, was first examined by Shackleton and Wild and later by three of Scott's four-man parties as they headed to and from the route to the Pole, as described more fully shortly.

By this time little attention was paid to the dolerites, and efforts concentrated on geomorphology or physiography, mainly by Taylor, and finding fossils in the Beacon Sandstones, by Debenham. In the Spring of 1911, Taylor carefully examined and surveyed Taylor Valley and 125 km of coastline from New Harbor northward. In the following summer (1912) he spent three months at Granite Harbor making similar studies, including going inland over the Mackay Glacier. He made remarkable maps of these regions, especially of the Kukri Hills–Taylor Valley area (Figure 2.29); nevertheless, the form of Solitary Rocks and its connection via Pearse Valley was still not accurately depicted, as was criticized by Shackleton's geologists in 1909. Taylor also produced a splendid monograph on the physiography of the McMurdo region as part of the final reports of this expedition (Taylor, 1922), as well as an inspired memoir on the overall expedition (Taylor, 1916).

Priestley went with the Northern Party spending the first season at Cape Adare where, like Borchgrevink, they were hemmed in and did limited geology. The following season they moved southward along the coast to the region of Mt. Nansen. They lived through their own epic battles of survival against long odds.

In the rocks found on the sledge near the tent where Scott, Wilson, and Bowers were found in November of 1912, some nine miles from One Ton Depot, were some coal containing fossils of the Permian gymnosperm *Glossopteris* (Seward, 1914), which became essentially an index fossil linking the southern continents together as Gondwana prior to dispersal. There has also been some speculation that fossils of the Southern Beech *Nothofagus*, which had been identified earlier by the Swedish Antarctic Expedition (1901–1904) on the Antarctic Peninsula, may also have been identified here by Wilson in his journal, but no specimen was carried back (Chaloner & Kenrick, 2015; and correction, 2016).

The full scientific reports on the expedition started coming out in 1914 and continued on until 1964 when the full collection of two volumes was published by the British Museum (see Smith, 1924 for access to all these reports). They are detailed and extensive, covering planetology and a great deal on the geology of the entire region from Cape Adare, where the Northern Party went, to the Beardmore moraine rocks found on Scott's sledge. Smith gives thorough reports on the metamorphic rocks (1921) of Cape Adare and the McMurdo Region, plutonic and

Figure 2.29 Detailed physiographic–topographic map of the greater Kukri Hills, Ferrar Glacier region by Taylor (1916, 1922); the dashed lines indicate the sledging journeys. In spite of the general high quality and attention to detail, the depiction of Solitary Rocks, which is easily accessible, is not particularly well done.

hypabyssal rocks (1924) of South Victoria Land, and volcanic rocks (1954–1963) of the Ross Archipelago, Cape Adare, Terra Nova Bay, and Granite Harbor, as well as a summary of the geology of South Victoria Land (1963), which is particularly well done. Debenham (1921) reports on the Beacon sandstones of the McMurdo and Beardmore areas, and Rastall and Priestley (1921) report on the sediments of Robertson Bay. As far as reports on the dolerites, not much beyond what is reported by Ferrar, Prior, Benson, and Browne, as already mentioned, was added by Smith (1924, p. 198), although some interesting field relations between the dolerites and the Beacon sediments are described by Debenham.

Needless to say, to work on the rocks collected by Wilson and carried on Scott's sledge to the end produced a certain poignant sentiment:

No one who has the privilege of examining these specimens and who knows the details of the tragic journey on which they were collected can fail to feel that here are specimens of inestimable value to be treasured in the national collections and to be examined with care and patience, so that, if possible, every stone may play its part in the piecing together of the geological history of the Beardmore Glacier. *(Smith, 1924, p. 219)*

### 2.2.5 Shackleton's Ross Sea Party: His Forgotten Men

Although often overlooked, as a part of Ernest Shackleton's epic *Endurance* or Imperial Trans-Antarctic Expedition of 1914–1915, a second party of men was landed on Ross Island in McMurdo Sound to support his cross-continental trek. These men were to lay out a series of supply depots across the Ross Ice Shelf to the Beardmore Glacier. Shackleton's own party aboard *Endurance* was to disembark in the Weddell Sea and traverse the continent, across the South Pole, descend the Beardmore Glacier, and head onward to Ross Island. Even though *Endurance* reached within 50 km of the coast, Shackleton never launched the mission and *Endurance* went on to be crushed by sea ice, and the team, after a long wait on the drifting pack ice, migrated via lifeboats to Elephant Island near the Antarctic Peninsula. Knowing they would never be found in this inconspicuous and hostile ice-covered island, Shackleton and a crew of five made the harrowing voyage aboard a roofed over lifeboat some 1,300 km to the island of South Georgia and a Norwegian whaling station. Landing on the wrong side, then crossing the formidable mountains, under much hardship they reached the station, but it took several tries to finally get a proper ship from Chile to get through the pack ice to reach his men on Elephant Island (see e.g., Lansing, 1959; Alexander, 1998).

Meanwhile the Ross Sea party was offloaded from *Aurora* at Cape Evans, Scott's *Terra Nova* base in January 1915. The ship under Captain Aeneas Mackintosh was to winter over, but a major gale in early May ripped *Aurora* from its anchors at Cape Evans (the anchors are still there), embedding it in pack ice where with a damaged rudder it drifted for 312 days before making port in New Zealand; Aeneas Mackintosh and nine others had been left at Ross Island. Although a goodly portion of food supplies had been offloaded, much of the expeditionary gear was still on board, and to carry out the depot-laying, sleeping bags, parkas, sledges, and other gear had to be improvised from items left at Cape Evans from Scott's last expedition. Nevertheless, under huge hardship, working partly in the polar darkness, they laid depots 70 miles apart containing 4,000 lbs. of provisions as far South as 83°37' near Mt. Hope at the foot of the Beardmore Glacier. Three men were lost: two (Aeneas Mackintosh and Victor Hayward) on drifting ice in McMurdo, and one (Arnold Spencer-Smith) due to exposure and malnutrition returning from the Beardmore. Just as they were getting ready to withstand another winter, butchering seals, on 10 January 1917, *Aurora* arrived carrying Shackleton, who had traversed the world from Chile and headed back to McMurdo (e.g., Bickel, 2001; Tyler-Lewis, 2006).

Anecdotal evidence suggests Ernest Joyce (who had been on three previous expeditions with Scott, Shackleton, and Mawson), Alexander Stevens, chief scientist, Richard Richards, physicist, Andrew Jack, physicist, and John Cope, biologist and surgeon, made cursory examinations of the rocks on the Beardmore, at Mt. Erebus, and Hut Point. Jack kept meticulous diaries (four volumes in the State Library of Victoria) with photographs and sketches, including some descriptions of Mt. Erebus eruptions (e.g., 12 March 1916). But apparently no systematic sampling was made, and the Dry Valley region was not explored.

After being busily occupied almost continuously since 1902, as *Aurora* sailed northward with Ernest Shackleton and the Ross Sea party aboard, McMurdo Sound, Ross Island, Hut Point, and the Dry Valley region again went silent to human activity as it had been for millions of years.

### 2.2.6 Closing of the Early Discovery Era: No Wright Valley

Every expedition entering the Transantarctic Mountains in this early era could not help but be impressed by the beautiful Beacon Sandstones and the ever-present magnificent sills of dolerite. All the way from King George Land, where Douglas Mawson had landed with his Australasian Antarctic Expedition (1911–1914), on the far north through Victoria Land to the top of the Beardmore Glacier, a distance of over 1,000 km, this association was readily found. All succceding expeditions throughout the Trans-Antarctic Mountains have been similarly impressed by the thick dark bands of dolerite sills within massive basal expanses of pink granite and extensive upper sections of tan sandstone. These dolerites reflect, as they do throughout the world, the dramatic Gondwana breakup at ~200 Ma, and the Dry Valleys apparently form the relatively un-faulted and un-deformed west shoulder of a rift flank (described in more detail in Chapter 13).

Throughout these early times of exploration, nevertheless, it is also remarkable that of the large expanse known today as the McMurdo Dry Valleys only the Ferrar-Kukri Hills-Taylor Valley corridor was discovered. And this discovery was essentially by accident, as Albert Armitage, frustrated at being trapped at Hut Point, tried to force a way through the imposing Royal Society Range. In all the careful mapping along the coast from New Harbor to Granite Harbor, traveling up and down past Marble Point, no one tried to do the same and force a way inward. They did sight and name Mt. Newell at the foot of Wright Valley, and if an ascent had been attempted what a glorious sight would have been discovered. This lack of discovering the obvious reflects the imposing forms of the Piedmont Glaciers along the west coast of McMurdo Sound. These glaciers essentially form a local coastal mountain range fed by snow falls and humidity precipitated by the close proximity of the ocean and often open waters of McMurdo Sound. Standing at Marble Point, looking westward the sharp ice edge and wall is formidable and uninviting. Even at the outlet of Taylor Valley, ice forms a massive barrier to easily navigating westward into the interior. Trips to Taylor Valley were always made by going up the Ferrar Glacier and around the snout of the Kukri Hills and then eastward into the valley, as Scott originally did. The discovery of the rest of the Dry Valleys needed the era of aviation and a single fly over. Throughout these times, this region was called "The Dry Valley."

Expeditions by Richard E. Byrd (1888–1957) in 1928–1930 and 1933–1935 were centered at Little America, near the Bay of Whales area on the Ross Ice Shelf about 700 km from McMurdo. Roald Amundsen (1872–1928) had staged his successful trip to the South Pole from here, arriving in December of 1911, about a month before Scott's team arrived. With aircraft, Byrd explored much of this region and geological work was carried out in the Queen Maud Range, finding there the Beacon sediments similarly intruded by a series of dolerite sills (Gould, 1935), but no work was done in the McMurdo region.

### 2.2.7 *The International Geophysical Year Era*

After the early discovery years, McMurdo was not visited in any detail again until the initiation of the International Geophysical Year (IGY) in 1957–1958 and establishment of the permanent bases of Scott Base (NZ) and McMurdo Station (US) on Ross Island. A prime idea of the IGY was to be able to make simultaneous geophysical measurements, like magnetospheric activity, at both ends of Earth. This furnished major opportunities for an abundance of scientific activity, including the first traverse of the Antarctic Continent by Vivian Fuchs, traveling from the Weddell Sea, and Edmund Hillary, traveling from McMurdo using Massey-Ferguson Tractors, meeting at the Pole in the Commonwealth Trans-Antarctic Expedition (1955–1958).

Although the sequence of discovery is not exactly clear, the first aircraft immediately saw the much larger region of dry valleys to the north of the Asgard Mountains. This prompted the first large scale systematic mapping within the full Dry Valleys by geologists Bernard Gunn and Guyon Warren, completing a remarkable ~1,500 km dog-sled journey circumnavigating the entire Dry Valleys. Going north along the McMurdo coastline and piedmont glaciers, they headed inland via the Mawson Glacier, and then traversed on the Polar Plateau southerly along the western edge of the Dry Valleys. They made many side trips deep into the valleys, going everywhere sleds could go, including along Finger Mountain, to Solitary Rocks and the west end of the Kukri Hills negotiating the major crevasse fields near Depot Nunatak (pers. comm. 1992–2000). They then continued southerly along the Polar Plateau, finally descending the Skelton Glacier and returning to McMurdo. The resulting 157-page research monograph and 1:250,000 scale geological map (Gunn & Warren, 1962) set the stage for all subsequent fieldwork, including the recognition of the overall internal structure of the sills and especially the "rhythmic banding" and dominance of massive amounts of orthopyroxene in the Basement Sill. Gunn's recognition of the importance of these geologic and petrologic relations via his subsequent publications form the basis, beginning in 1992 until 2008, of the present field study (i.e., Gunn, 1962, 1963, 1966). More specifically, in Gunn (1966) many sampling profiles were presented along with detailed chemical, petrographic, and petrologic analyses, which revealed the fundamental value of this magmatic system, aside from any systematic order between the various sills.

In late 1958 Warren Hamilton of the USGS made a detailed study of mainly the Basement Sill in the Upper Taylor Glacier region (Hamilton, 1964, 1965), including detailed stratigraphic sampling and chemical and petrographic analysis of a section in the south wall of Pearse Valley, which he called Solitary Valley. Moreover, in a perceptive aerial reconnaissance he realized the large extent and volume (over 100 cubic miles) of the Basement Sill and noted its thinning and truncation in the north wall of eastern Wright Valley.

Some of the more useful comprehensive mapping, exposing the systematic distribution of the dolerites was due to extensive mapping on foot starting in 1958 by students from Victoria University in Wellington; the so-called Victoria University Antarctic Expeditions (VUWAE). This work was initiated by Barrie McKelvey and Peter-Noel Webb in the upper Taylor Glacier area and continued the following year with a larger group (including Colin Bull) mapping in Wright and Victoria Dry Valleys (e.g., McKelvey & Webb, 1959, 1962; Webb & McKelvey, 1959). A particularly engaging historical account of these times is given by Bull (2009).[10]

A great deal of additional valuable general mapping, including the geomorphology and glacial history, has been, more or less, continuously going on by various groups, especially the New Zealand Geological Survey, and this vast assortment of results has been skillfully collected and elegantly presented by Cox et al. (2012; also, Turnbull et al., 1994). Some of these works do involve the dolerites, but mainly only as cursory or specialty studies involving characterizing

---

[10] Following an obscure contact high along the east wall of Bull Pass, in 2008 the author found a well-used blue handled geology hammer. Judging from the severe UV damage, it is most probably from this era of mapping, although Colin Bull had no recollection of loss of a hammer by anyone (pers. comm. 2010).

their isotopic identity or exact age. My initial work in the 1992–1993 field season at Solitary Rocks, for example, concerned the origin of the silicic segregations or lenses found by Gunn (1966) in the upper parts of many sills (e.g., Marsh, 2002). The prevailing impression at the time was that the dolerite system was complex and perhaps haphazard, both geochemically and geologically. It was only with the recognition of the dynamic meaning of the regionally pervasive tongue of orthopyroxene that the delicate systematic collective order of this entire magmatic system became apparent. Nine field seasons were spent on these rocks, with camps at Solitary Rocks (2) and Bull Pass (7). This included a "Magmatic Field Laboratory Workshop" in 2005, attended by 25 petrologists, based in Bull Pass, with detailed studies throughout the Dry Valleys, including thin section-making and petrographic work in McMurdo (Bedard, 2005; Jerram, 2005; Mathez, 2005). Although ostensibly centered on the Transantarctic Mountains, the treatise by Faure and Mensing (2010) is a major resource on most aspects of Antarctic research.

## 2.3 Physiography

The Transantarctic Mountains, in essence, dam the northeasterly flow of ice forming the polar plateau or East Antarctic Ice Sheet. The surface elevation of the 1–3 km thick ice sheet is 2.5–3.5 km to the southwest toward the polar interior and drops to near sea level where it breaches and traverses the Transantarctic Mountains, forming the massive Ross Ice Shelf. These breaches occupy a series of generally east–west trending major valleys in the McDV, which are bounded in the N–S directions by smaller mountain ranges (Figures 2.30 and 2.31). The general form of these inter-valley mountains (e.g., Asgard and Olympus ranges) is a series of upland plateaus supporting high buttes of Beacon sandstone often capped or internally trussed by dolerite. The original geomorphic form is due primarily to water and wind erosion in the early Tertiary before establishment of the ice sheet, which began in the Oligocene (~30 Ma ago) when Antarctica disconnected from South America, allowing development of the circumpolar current (Denton et al., 1993). The terrain is, for the most part, an ancient or relict terrain, which is substantiated by cosmic ray exposure dating (e.g., Brook et al., 1995) and began uplifting at ~55 Ma ago (Fitzgerald, 1992). Further small-scale sculpting of this basic landform has been due to intermittent alpine-style glaciation (Sugden & Denton, 2004; Denton & Sugden, 2005; Lewis et al., 2007). The detailed erosional history, the history of glaciation, and even the long-term stability of the east Antarctic ice sheet has been debated in some detail, which is more than interesting; for useful reviews see Faure and Mensing (2010) and Cox et al. (2012).

Figure 2.30 Scenic views of the Ferrar dolerite sills in the McMurdo Dry Valleys. (a) Solitary Rocks in the center with Basement and Peneplain Sills, and Finger Mountain in the far-right corner. (b) Upper Wright Valley with upper contact of the Basement Sill, Peneplain Sill at the Labyrinth, Asgard Sill, and Mt. Fleming Sill in the distance. (c) Bull Bass (looking north) with thick Basement Sill on the left and thinning greatly to the East (right). (d) Uppermost Wright Valley with Asgard Sill structure and Mt. Fleming in the distance.

The McDV are starved of ice due in a large part to the erosional resistance of the massive complex of dolerite sills, which has limited the extent of penetration by breaching ice flows from the polar plateau (see Figures 2.4, 2.30, and 2.31). The northeastward flow of ice is deflected around these valleys, which are

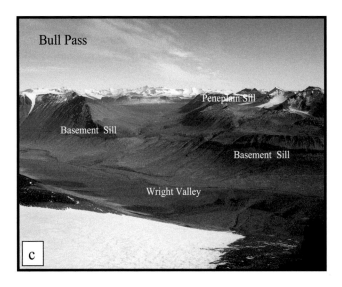

Figure 2.30 (*cont.*)

themselves dehydrated by the desiccating effect of adiabatic heating (by about 40°C) by exceedingly dry katabatic winds cascading from the polar plateau through the valleys to near sea level. It is the combination of these conditions (i.e., massive dolerite and drying winds) that has largely preserved the Dry Valleys terrain, which, judging from the delicateness of the high-country buttes and the forms of the valleys, has evidently never experienced full-scale continental-style glaciation. The effects of long-term wind erosion are prominent

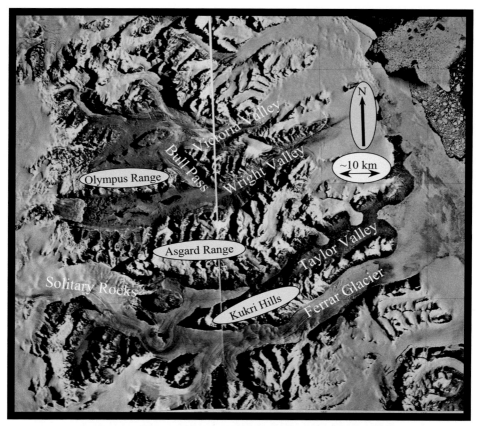

Figure 2.31 Topographic relief map of the McMurdo Dry Valleys (from US Geological Survey)

in common sand dunes and outcrops tortuously sculpted on all scales from millimeters to meters. In addition, Wright and Victoria valleys have been swept uncharacteristically free of erosional debris by the action of torrential floods from the release of enormous quantities of water trapped within and transported by the interior ice sheet (Denton et al., 1993). Enormous vats or lakes of water contained and transported within the ice cap have been dumped from high elevations at the heads of Wright and Victoria Valleys, sweeping these valleys clean of rock debris; mega ripples (1–2 m high) cut in granite country rock are found at the east end of Lake Vanda in Wright Valley, as are huge "thunder pits" in central Bull Pass. Similarly, massive subglacial outwash at the head of Wright Valley has deeply incised the Peneplain Sill forming the anastomosing canyon terrain known as the Labyrinth (Lewis et al., 2007). The estimated flux of water involved is ~$2 \times 10^6$ m$^3$/s with the last major flood event occurring in the middle

Miocene. This phenomenon has provided for possibly the cleanest, most revealing field relations found anywhere on Earth.

## 2.4 General Geology

An extensive, sharp, and low relief (<~10 m over 5–10 km) Devonian erosion surface, the Kukri Peneplain or Erosion Surface, divides the entire region into two principal groups of rock, a thick upper section of sandstones called the Beacon Supergroup, and a basement complex of mostly granitic plutonic rocks called the Granite Harbor Intrusives (Figure 2.32). The Ferrar dolerite sills mainly intrude the Beacon sandstone except for the lowermost sill, the Basement Sill, which is almost always just below the Kukri surface in the plutonic rocks.

The Beacon Supergroup is a generally flat lying sequence up to 2.5 km thick resting unconformably on Ordovician and older rocks throughout the Transantarctic Mountains (Barrett, 1991). It comprises two groups: A quartzose sandstone sequence of Devonian age called the Taylor Group, and a heterogeneous sequence of Late Carboniferous to Early Jurassic age called the Victoria Group. In the McDV region where it was first seen and described by Ferrar (1907) the Beacon consists of a heterogeneous sequence of siltstones, sandstones, coal measures, and conglomerates from Devonian to Jurassic in age (McKelvey et al., 1970, 1977). Concise and useful descriptions of these units in this area are given by McElroy and Rose (1987) and Cox et al. (2012).

The granitic plutons of the Granite Harbor Intrusives, which are below the Kukri surface, are Pre-Devonian in age, of a wide range of sizes, shapes, textures, and of compositions mainly of granite, granodiorite, and quartz monzonite. They are pink, gray, and brown, medium-to-coarse grained, and often porphyritic in feldspar. Detailed mapping of crosscutting relationships and reconnaissance dating suggests that much of this complex is Ordovician in age (Allibone et al., 1991; Cox et al., 2012). These plutons are bordered on the east, toward McMurdo Sound and mainly outside the area of the dolerites, by a series of Proterozoic metasediments (marbles, schists, and gneisses) called the Koettlitz Group (Grindley & Warren, 1964).

The pre-Kukri basement rocks are cut by several sets of dike swarms, older mafic and felsic, and younger dolerite. In the east Wright Valley, northeasterly striking mafic and felsic dikes pervasively cut the basement complex. They are generally 1–5 m wide, are commonly porphyritic in feldspar, and the felsic dikes sometimes cut the mafic dikes. Although not obvious in many areas, their pervasiveness is clear from their occurrence throughout Wright Valley, in the higher north wall of Taylor Valley, and at Cavendish Rocks near Solitary Rocks at the western end of Taylor Valley. Although appearing in the field as a felsic and

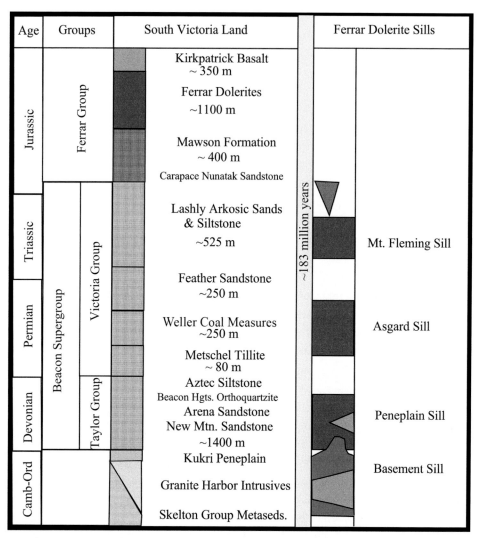

Figure 2.32 Simplified stratigraphic column for the McMurdo Dry Valleys area; partly after Barrett (1991)

mafic bi-modal population, a detailed study of some 600 of these dikes in Wright valley shows them to form an almost continuous compositional population from 48 to 78 wt.% $SiO_2$, and with a commensurate range in radiogenic isotopes with initial $^{87}Sr/^{86}Sr$ of 0.7118–0.7085 and $\varepsilon_{Nd}$ from −12.7 to 8.0 (age corrected; Bray et al., 2009; Cox et al., 2012). There is also no clear correlation between bulk composition and isotopic composition.

Younger dolerite dikes cutting the Beacon sediments, associated generally with the emplacement of the Ferrar dolerite sills, but not specifically associated locally with any local sill, are also found in some places; for example, in the upper valley walls at Solitary Rocks and within the Asgard and Olympus Ranges. These "regional" dikes are near vertical, generally strike approximately E–W, are 1–5 m in width, and are almost always cut by the sills themselves. Relative to the amount of magma associated with the dolerite sills, these dikes are scarce and in the whole of the Asgard and Olympus Ranges there may not be more than a dozen in each range. A local dike swarm is well exposed associated with a basal sill at Terra Cotta Mountain, directly south of Solitary Rocks, which is unusual (Morrison & Reay, 1995). A broadly similar fanning dike swarm is also found about 150 km to the north in the Clare Range. A swarm of apparently similar dikes is exceptionally well exposed in the granites of the upper east wall of central Bull Pass, but from their degree of alteration and composition they are clearly not of Ferrar age. There are also countless small dikes emanating from the contacts of the individual sills, which are associated with the detailed emplacement processes; more of which will be said when discussing emplacement processes.

The Ferrar dolerite sills cut all units mentioned so far (Figure 2.33). Rare dolerite dikes that cut the sills have been reported, but the author has not observed

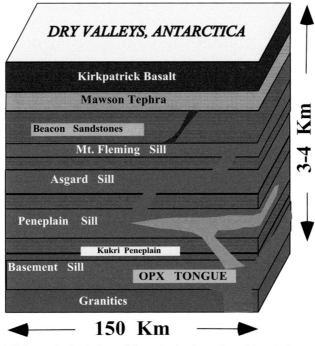

Figure 2.33 Schematic depiction of the principal stratigraphic relations and general nature of the Ferrar dolerite sills, including the placement of the strong concentration of orthopyroxene primocrysts (Opx Tongue) in the Basement and Peneplain Sill (revised after Marsh, 2015)

these. In the general area of Wright and Taylor Valleys, as described at the outset of this section, there are four major sills over a vertical height of about 3.5 km. The lower two sills (Basement and Peneplain Sills) are the thickest, each ~350 m, and the upper sills (Asgard and Mt. Fleming Sills) are 250–300 m thick. Although the lower sills seem more expansive, cropping out for over 100 km south to north, the upper sills, which today are erosional remnants, may originally also have been similarly expansive.

To the north, in the area of Allan Hills, Carapace Nunatak, and Coombs Hills, which border the main focus area of the present study, small sill- and dike-like bodies can be traced almost continuously upward throughout the transition from dolerite to tephra (Grapes et al., 1974; Ross et al., 2008). This tephra forms an extensive subaerial phreatomagmatic deposit known as the Mawson Formation, which is up to 170 m thick (e.g., Gunn & Warren, 1962; Korsch, 1984; Elliot & Hanson, 2001). This formation is capped by locally extensive flood basalts, the Kirkpatrick Basalt, which are the volcanic equivalent of the upper dolerite sills (e.g., Fleming et al., 1995). The nature of the eruptive environment during Ferrar magmatism time was characterized in this region by central complexes of nested diatremes forming local depressions containing rafts of country rock and volcaniclastic material in the form of tuff breccias and lapilli tuffs with "swirly" thin basaltic dikes "tangled" within (Ross et al., 2008). This central vent structure or localized magmatism is consistent with the weak formation of local dike swarms and also with the basic emplacement pattern of the massive sills themselves. The highest density of dikes instead seems to be associated with high-level sill emplacement, as mapped on Mt. Fleming by Pyne (1984). The basic layer-cake sill and sediment stratigraphic structure, so common at deeper levels, also holds at Mt. Fleming, but with a more pronounced tendency for the sill component to be self-faulted due to emplacement dynamics and with the principal sill to breakup into associated much smaller, coeval sills (Figure 2.30c). At higher levels, on the order of 500 m higher, as 20 km north at Shapeless Mountain (2,736 m), the layer-cake structure is gone (hence the name Shapeless Mtn.) and there is a strong sense of a transition from the deeper sill structure to a near-surface volcanic environment, which Korsch (1984) shows well in his smaller scale mapping (e.g., his figure 1). Tilted and fragmented massive blocks of Beacon sediments are bathed in central areas of dolerite, and dolerite dikes grade into basalt and then into Mawson Formation breccias. Fluvial processes have reworked Mawson Formation debris that has been subsequently intruded by still younger Mawson Formation breccia, and there are abundant signs of local phreatic explosions.

The fundamental character of the Ferrar volcanism in these high altitude, outlying areas is of central vents of phreatic eruptions emanating from near surface sills, very much as sills punching vents through a relatively thin overlying brittle lid of Beacon (e.g., Ross et al., 2008). The role of this style of magmatism–volcanism in establishing the Ferrar McDV magmatic plumbing system, and the Ferrar magmatic province in general will be visited in some detail in Chapters 12 and 13.

The youngest igneous units in this region are a series of alkali basalt cinder cones with lavas, feeder dikes, and scoria 1.5–4 Ma in age (Armstrong, 1978; Denton et al., 1993; Wilch et al., 1993). They are part of the Cenozoic McMurdo Volcanic Group and are associated with the extensive volcanism on nearby Ross Island, which defines the Erebus Volcanic Province (Kyle & Cole, 1974; Wright & Kyle, 1990). These scattered volcanics drape over the ancient topography and mainly occur on the margins of Taylor and Wright valleys and on the Kukri Hills. Although deeply dissected, excellent exposures can be found of lavas venting from dikes and sequences of lavas and tephra, containing lapilli and bombs, and small pyroclastic flows, all of which is suggestive of subaerial Strombolian activity (e.g., Allibone et al., 1991; Wilch et al., 1993). The lavas themselves are vesicular olivine basanites with micro-phenocrysts of olivine and clinopyroxene and microlites of plagioclase. What may be even more novel is that these eruptions have clearly taken place on the already well-sculpted Dry Valleys topography. Vents are found on a variety of topographic extremes from cirque ridges to small plateaus underlain by Ferrar dolerite sills, clearly exposed in ancient cuts nearby, and to vents on steep mountain or valley walls. And when vents overlie in close subsurface proximity to Ferrar sills there are often xenoliths of Ferrar dolerite in the scoria ejecta showing the effects in various stages of reheating and attempted assimilation. The distinct paucity of these vents on the valley floors and on the inner or highest parts of the mountain ranges reflects the control of topographic stresses in directing the upward propagating dikes, as is also seen in the western United States.

What is also remarkable is the persistence of this basic geological framework throughout the Ferrar magmatic system of the Transantarctic Mountains (TAM). A series of four major sills, with sundry lesser local sills, are found almost everywhere at similar stratigraphic levels and of broadly similar compositions, which may reflect the rapid development and abandonment of this Gondwana-rift related magmatism (e.g., Elliot, 2013; Elliot & Fleming, 2017).

# 3

# The Ferrar Dolerite Sills

## 3.1 Introduction

The overall nature, style of occurrence, age, and basic petrologic structure of the sills as a group are first described and then the individual sills are discussed in Section 3.2. Since in much of the TAM the deeper parts of the Ferrar are poorly exposed, it is useful here to describe the details of the sill exposures with the intent that they may be of general relevance throughout the range.

## 3.2 Occurrence

As introduced in Chapter 2, four massive sills occur throughout the region (~10,000 km$^2$, Figures 2.30, 2.33, and 3.1). From the bottom up, these are the Basement Sill (350–500+ m), the Peneplain Sill (350 m), the Asgard Sill (300 m), and at the highest elevations there is often a fourth sill, which is here referred to as the Mt. Fleming Sill (100 m). As the names suggest, the Basement Sill almost always occurs below the Kukri Peneplain in the granitic plutons, the Peneplain Sill generally just above the Kukri Peneplain in the lowermost Beacon sandstones, and the Asgard Sill forms the summits of the western Asgard Mountains and fragments cap the Inland Forts well up in the Beacon. The Mt. Fleming Sill is at the topmost section of the Beacon, just below the transition to tephras and lavas. The Mt. Fleming Sill is typically intruded into the Late Permian Weller Coal Measures or Feather Conglomerate (Pyne, 1984), whereas the Asgard (Beacon) Sill is typically intruded into the Late Devonian Beacon Heights Orthoquartzite (McElroy & Rose, 1987).

Although the sills are generally remarkably free of structural complexity, they are sometimes irregular in form and stratigraphically discontinuous in certain places. These irregularities are most prominent in the Asgard Sill and Mt. Fleming Sill as humps and sharp, fault-like discontinuities (e.g., Figure 2.30c).

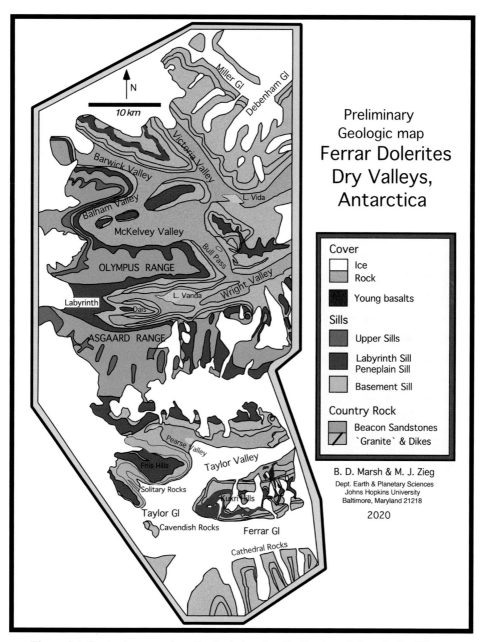

Figure 3.1 Geologic map primarily depicting the distribution of the principal Ferrar dolerite sills in the McMurdo Dry Valleys (revised after Marsh, 2015)

The Peneplain Sill has the truest form and is for the most part a sheet of uniform thickness throughout the region. Outside Wright Valley, the Basement Sill is also highly regular and uniform and may be the most extensive of all, but within Wright Valley it has a lobate structure in Bull Pass and especially the south wall of Wright

Valley. Here it forms three large, distinctive, and interconnected lobes near the Comrow Glacier and Mt. Odlin. The lower contact of these lobes climbs upward along gently dipping pre-Kukri dikes and the upper contacts generally follow until quenching against the overlying Peneplain Sill. Beneath Mt. Odlin the upper contact appears faulted, but this actually reflects splaying in the propagating edge of the sill during emplacement. Clear examples of this are also displayed by the Asgard Sill where it forms the Airdevronsix Icefalls at the head of Wright Valley and also nearby in the Mt. Fleming Sill (e.g., Pyne, 1984). The presence of these features seems to reflect the direction of emplacement, which would have been in the direction of strike of the apparent fault. The critical evidence is the presence of continuous chilled margins against and along the apparent offsets, as opposed to offsets in the chilled margins themselves.

### 3.3 Style

Viewed as a sequence of sills upwards through 3–4 km of crust, it is as if at one time the entire overlying crust was floating on massive sills. The sills are, in general, remarkably flat lying with well over 90 percent knife sharp, strongly chilled margins. (There is a clear secondary regional tilt to the whole sequence of 2–3° to the west due to post-rifting tectonic adjustments (Denton et al., 1993).) Nevertheless, blocks of Beacon sandstone can also be seen in some places (e.g., Finger Mtn., Friis Hills, SW Upper Wright Valley) falling from the upper contacts of the Peneplain and Asgard Sills, with some blocks descending 50 m or more into the sill. The blocks always maintain their integrity, although there is a certain amount of fragmentation around the edges, with chilled dolerite around all margins. No blocks have been seen as deep as midsection in the sills. There are also some restricted areas in the Basement Sill where the upper contact is not sharp but is remarkably diffuse or "digested," as noted by Gunn (1966). These are areas where the dolerite locally penetrates or climbs upward, apparently partially melting, disaggregating, and sometimes ingesting the granitic roof rock, the debris of which has mixed with the dolerite (Figure 3.2). These are henceforth called *climbing contacts*. This dolerite is often distinct in the conspicuous presence of large plagioclase phenocrysts inherited from the granitic rock, dispersed downward by as far as 10–20 m and also in the presence of older internal chilled margins that have been overrun as the dolerite continued to climb upward. The dolerite in these areas also does not form a strongly chilled margin even at the leading edge of the last contact but is of a medium-fine grain size. Places of distinctive Basement Sill climbing contacts are at the northeast end of the Dais in Wright Valley, which forms the roof of the Dais Intrusion, the southwest corner of Bull Pass, north–northeast of the east end of Lake Vida, and most pervasive of all the entire contact

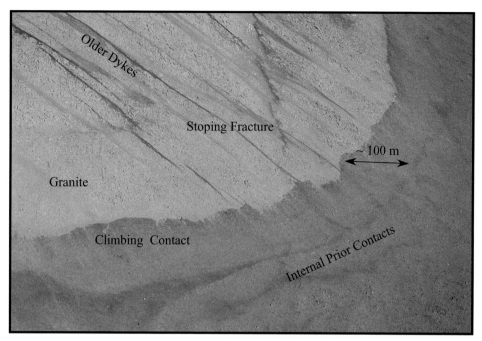

Figure 3.2 Typical climbing contacts of Basement Sill in Central Bull Pass, which is similar all along this contact, extending northward and easterly into Victoria Valley. Also notice the distinct incipient stoping defined by the fracture cutting diagonally upward from the sill into the granite.

from the east central wall of Bull Pass, continuing north into Victoria valley and then eastward, a distance covering about 8 km (Figure 3.1). At each of these locations the overlying granitic country rock is pervasively melted and at the latter-most location the melting is extensive, extending some 100 m roof-ward. This process and the implications of this melting will be discussed in much more detail in Chapter 6.

## 3.4 Age

The Ferrar dolerites throughout the Trans-Antarctic Mountains mark the Gondwana breakup event at about 182.7 Ma, with the overall time of full system development being on the order of the remarkably short time of ~0.4 Ma (e.g., Heimann et al., 1994; Fleming et al., 1997; Elliot & Fleming, 2017; Ivanov et al., 2017).

Of more paramount importance here is the relative timing between emplacements of the various sills, which bears on the general sequential development of the entire system; that is, bottom up or top down. Although the Basement and Peneplain sills are often separated by 100–200 m, in a number of places they touch (e.g., west end of Kukri Hills, Friis Hills, and several places in the south wall of Wright Valley). In all cases the Basement Sill is quenched against the Peneplain Sill, which was also noted by Gunn (1966) at the west end of the Kukri Hills and on the south side of the Friis Hills. That the Peneplain Sill was not yet solid when the Basement Sill was emplaced, however, can also be seen at the Friis Hills where the Basement Sill has pushed into the Peneplain Sill and formed a "soft" chilled margin. Because the solidification time of a 350 m thick sill is about 1,000 years, the time between emplacements of the two sills was thus short, perhaps only about 100 years. Perhaps even more telling is the continuation of the Opx Tongue, so characteristic of the Basement Sill, that is also found in the Peneplain Sill west of central Bull Pass, continuing west to Marsh Cirque and farther west in the Labyrinth at the head of Wright Valley. That the Peneplain Sill was emplaced early may also be reflected in its generally flat, uncomplicated, and areally pervasive nature. After some solidification its presence stiffened the sedimentary section of the crust, making succeeding sill-forming injections adhere to a highly ordered neutral stress field, promoting horizontality. And the close temporal (and spatial) proximity of the emplacement of the Basement Sill may be reflected in the occasional digested upper contacts, perhaps reflecting wall rock still relatively hot from Peneplain Sill emplacement. Moreover, as will become apparent, the Upper Lobe of the Basement Sill beginning on the east wall of Bull Pass and extending through the Asgard Range to the Kukri Hills is softly quenched against the Peneplain Sill at Mt. Grendall (south wall of Wright Valley), clearly indicating that both magmas were mushy. This may also indicate that this Upper Lobe was actually emplaced before the main lower, principal lobe of the Basement Sill.

The relative ages of the remaining sill sequence are more difficult to discern from the limited field exposures available upward in the section. But judging from their relative degrees of alteration, the Mt. Fleming Sill, the uppermost sill, is older, and thus the sills increase in age in the sequence from bottom to top. The remarkably flat nature of the Peneplain Sill may reflect the prior presence of the upper sills in stiffening the crust against significant deformation by the later sills. That the upper sills are more irregular in form may reflect their early emplacement and their proximity to the surface, but once emplaced and mostly solidified they stiffened the crust below. The local irregularities near Bull Pass in the Basement Sill, over that of the Peneplain Sill, may reflect the hot, deformable nature of the crust at this level and later stage of the overall emplacement process.

Several lines of evidence, especially in the local nature of the schlieren in the Opx Tongue along with a small, layered intrusion, suggests a deeper massive feeder existing in central Bull Pass, from which emanates a stack of outward propagating lobes. The upper lobe of the Basement Sill in southeast Bull Pass, below Mt. Orestes, for example, continues southward through the Asgard Range across to the Kukri Hills, linking up with the lower, main lobe that forms the lower part of the Friis Hills and Solitary Rocks and the western nose of the Kukri Hills, there forming an otherwise unusual mushy contact. These more involved relations, it has proven, are best understood when introducing the detailed sill compositional profiles, especially as related to the Opx Tongue, which is a direct indicator of magma movement. This will all be presented in increasing contextual detail as the full sill system is presented in each succeeding chapter.

# 4

# Ferrar Basic Petrologic Structure

## 4.1 Stratigraphy

The sills are of two main petrologic types (Figure 4.1): Those that have little to no obvious internal structure (i.e., typical uniform dolerites and diabases common throughout Earth's Jurassic rift system) and those with distinct cumulate or rhythmic layering. The main reason for this difference is simply the absence or presence of large (1–20 mm) crystals of predominantly low calcium pyroxene (Opx) in the initial magma. Although also often containing 10–15 percent high calcium pyroxene (Cpx), this assemblage is hereafter referred to as orthopyroxene or Opx or the Opx Tongue. And due to their size, these crystals will, for convenience, be called phenocrysts and/or primocrysts, even though at a fundamental level they are clearly xenocrystic. This distinction cannot be emphasized too strongly, for it colors all petrological aspects of this system and may well be the most fundamental finding of this study. All dolerite containing Opx phenocrysts shows sorting and layering of some kind. All dolerite devoid of Opx phenocrysts is, to first order, uniform and featureless. This declaration is decidedly tempered by the detailed study of the apparently featureless Beacon Sill at Beacon Heights (SW of Solitary Rocks), showing a systematic process of multiple injections and subtle internal contacts (Zieg & Marsh, 2012). Through extensive field observations, petrography, and bulk chemical analysis, Gunn (1966) recognized these two basic sill types. He divided them into hypersthene-bearing sills and pigeonitic or non-hypersthene-bearing sills. He also recognized that the large Opx crystals did not grow in situ, but were contained in the initial magma, and that the internal structure was mainly due to these phenocrysts.

Aside from this modal and textural division, the only other major feature is the widespread presence of a distinctive horizon of coarse-grained silicic segregations in the upper parts of most sills (Figure 4.1). They first appear in the upper 20–30 m as small splotches (1–5 cm in diameter) and lens-like segregations (1 by 30 cm)

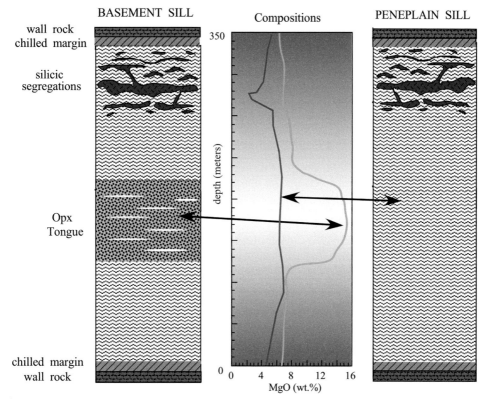

Figure 4.1 The characteristic petrological and chemical stratigraphy in the Basement and Peneplain Sills, illustrating the strong effect of the presence of the Opx Tongue in the Basement Sill, as originally pointed out by Gunn (1963)

and increase in size downward, culminating as large inter-digitating horizontal lenses 1–50 m long and 0.1–2 m thick that in some places connect vertically to nearby segregations through a system of tears or fractures. Segregations of this type, which are also referred to as pegmatitic or granophyric segregations, are common to dolerite sills and lava lakes and their nature and mechanics of formation have been discussed elsewhere (e.g., Marsh, 2002; Boudreau & Simon, 2007; Zavala et al., 2011). Here they are described and discussed only for completeness in relation to the overall petrology and solidification of the sill complex itself. (Although all of a broadly similar nature, those in the upper reaches of the Dais Intrusion, especially on the northeast edge, are distinctive in being much more globular and uniformly grained, as if they have been remobilized or made of included granitic melt.)

Silicic segregations in dolerite sills and lava lakes form in response to a gravitational instability in the downward thickening upper solidification front (Marsh, 1996, 2002). They mark a critical point where the downward propagating solidification front becomes sufficiently distended and heavy to tear apart at a location where the matrix is still weak enough to fail (Figure 4.2). (Imagine a hanging worn carpet where with enough weight hung at its lower edge a roughly horizontal tear will somewhere suddenly appear.) From their bulk chemical compositions, they form within an area of significant strength in the solidification front where the crystallinity is 60–70 vol.%. Segregations have knife-sharp upper contacts and diffuse lower contacts. The grain size of the constituent minerals, chiefly clinopyroxene and plagioclase granophyric with quartz, is usually coarse (cpx may be 10 cm long) at the top and grades downward into more or less medium-grained normal dolerite. The absence of chilled margins and the variation in grain size indicates filling of a tear by local interstitial melt multiply saturated in clinopyroxene and plagioclase. The melt at the upper contact of the segregation was thermally and chemically in equilibrium with the upper contact and enjoyed unhindered crystallization, but the lower melt was being withdrawn upward from the underlying solidification front and was increasingly in thermal disequilibrium, prompting crystallization at a progressively finer grain size. In some areas where sill contacts are inclined and complicated (as in the lobate areas), the pattern of segregations in also complex, possibly suggesting the disruption and transport of earlier formed segregations deeper in the system. Alternatively, since these irregular segregations are apparently confined to the Opx Tongue of the Basement Sill, they may represent crystal-free pockets segregated within grain-supported areas of Opx Tongue sludge. The absence of strong spatial variations in grain size within these segregations perhaps supports the latter interpretation.

## 4.2  Composition, Modal Mineralogy, and the Opx Tongue

The sills are ultramafic orthopyroxenite[1] to noritic in composition, with silica contents of 51–56 wt.% $SiO_2$ and 5–20 wt.% MgO. They contain plagioclase, orthopyroxene, and clinopyroxene as major phases (i.e., high calcium pyroxene with or without pigeonite); minor phases are quartz, alkali feldspar, and apatite, and there is an unusual paucity of Fe-Ti oxides. In the phenocryst-free sills (i.e., no Opx phenocrysts) the textures run from exceedingly fine-grained, ceramic-like, chilled margins to medium-grained holocrystalline, ophitic to subophitic, in the sill interiors, with pervasive splotches of granophyric intergrowths (quartz, alkali feldspar, and plagioclase). In the phenocryst-rich areas (i.e., Opx Tongue regions) there are always fine-grained phenocryst-free chilled margins (often also

---

[1] As will become readily apparent, the richly diverse petrographic character of these rocks makes it cumbersome to consistently employ standardized lithologic names such as websterites, gabbro-norites, leuco-gabbro-norite, etc., and instead more obvious names such as orthopyroxenites and anorthosites, etc. are used; but see also Bedard et al. (2007).

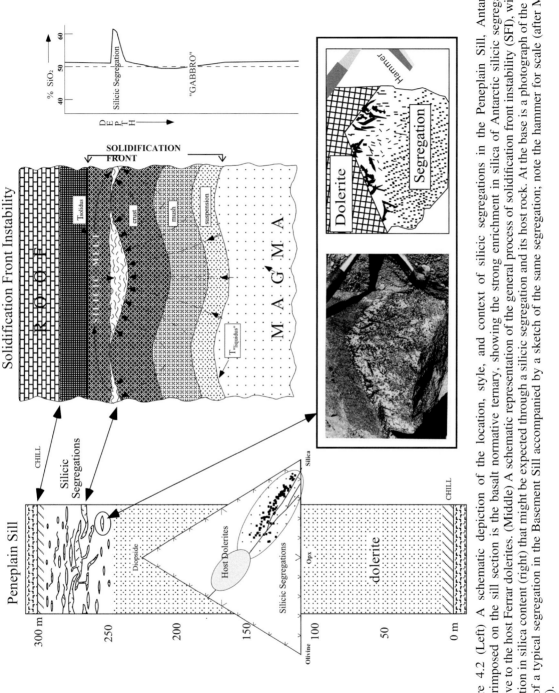

Figure 4.2 (Left) A schematic depiction of the location, style, and context of silicic segregations in the Peneplain Sill, Antarctica. Superimposed on the sill section is the basalt normative ternary, showing the strong enrichment in silica of Antarctic silicic segregations relative to the host Ferrar dolerites. (Middle) A schematic representation of the general process of solidification front instability (SFI), with the variation in silica content (right) that might be expected through a silicic segregation and its host rock. At the base is a photograph of the upper part of a typical segregation in the Basement Sill accompanied by a sketch of the same segregation; note the hammer for scale (after Marsh, 2002).

93

ceramic-like) grading strongly into fine grained dolerite and then into dense-packed, coarse-grained (1–20 mm) cumulates of orthopyroxene (and, always, associated subordinate clinopyroxene phenocrysts) and fine-grained (0.1–0.5 mm) plagioclase interspersed with stringers and layers (1–50 cm thick) of fine-grained plagioclase and orthopyroxene.

The Basement Sill contains, almost exclusively, the extensive, highly distinctive tongue of large orthopyroxene phenocrysts (i.e., the Opx Tongue hitherto and hereafter), which forms ubiquitous cumulate textures wherever the Opx Tongue exists. These cumulate textures include strongly sorted pyroxenes as well as anorthositic layers, bands, and stringers (sometimes hereafter collectively referred to as schlieren), along with attendant large systematic spatial variations in chemical composition. The Opx Tongue is most massive in Bull Pass and thins outward in all directions. In the area of filling, as already mentioned, near central Bull Pass, the Opx Tongue also extends upward and westward into the Peneplain Sill and is especially clear in the Olympus Range between Mt. Orestes, Mt. Jason, Marsh Cirque, and the Labyrinth. Weak layering or banding is clear at the Labyrinth, and the MgO content of this tongue is at most 17 wt.%. No similar diagnostic phenocrysts of Opx have been found higher in the system in any of the sills, dikes, or lavas, although much smaller, trace Opx cognate crystals are found in some lavas. The regional occurrence of Opx phenocrysts is an important diagnostic tracer of the filling dynamics.

The filling dynamics of the other, featureless sills are also fundamentally important to establish, but in not containing any obvious diagnostic concentration of large crystals, this is much more challenging to determine, which is the general situation for most dolerite sills worldwide. Yet, a careful study in the field often gives the impression of subtle and perhaps systematic variations in gran size throughout some sections of the Peneplain and Asgard sills. This impression led to a detailed examination of the Beacon Sill, in Beacon Valley, which is the equivalent of the Asgard Sill to the north (Zieg & Marsh, 2012). The sill here is 150 m thick with clear upper and lower contacts and full exposure in between; it was sampled at 5 m intervals and finer in some instances. These samples were analyzed for major and trace elements, and the CSDs (Crystal Size Distributions) were determined. Although this sill is remarkably uniform on a field and macroscopic scale, it displays distinct chemical and textural variability indicative of a sustained and complex history of emplacement and differentiation. The overall emplacement scenario is depicted in Figure 4.3, and a more generic depiction is given in Figure 4.4.

Emplacement of the sill consisted of at least two, and probably four, discrete massive injections of magma, each averaging 35 m thick, over a time span of ~100 years. The final injection event is marked by a 30 m thick interval of significantly

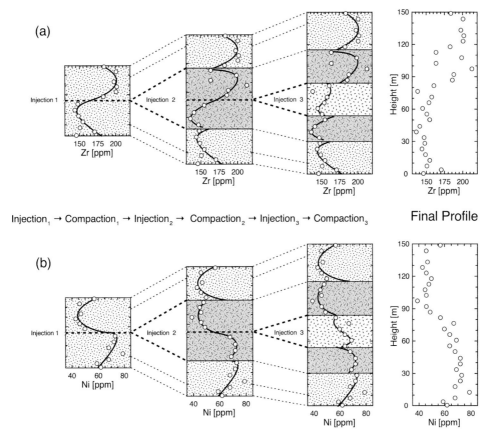

Figure 4.3 The probable sequence of emplacements forming the Beacon Sill in Beacon Valley, which is a typical nearly featureless sill. These two series depict the sequential development of the sill as recorded by the distributions of (a) an incompatible element Zr (zirconium) and (b) a compatible element Ni (nickel) (after Zieg & Marsh, 2012).

finer-grained rock at the sill center, where the earlier sill material was hottest. These fine-grained textures can be successfully reproduced by combining the results of thermal and crystal growth models, but only if the sill was a multiple intrusion. With the exception of a minor reinjection event captured in the chilled margin of the sill, earlier reinjection events are not texturally evident, most probably because of textural overprinting during prolonged cooling after initial crystallization. The dominant process involved in the post-emplacement differentiation of the Beacon Sill was compaction driven redistribution of interstitial liquid, as has been well documented by A. R. Philpotts and associates (e.g., Philpotts et al., 1996; Philpotts & Philpotts, 2005) in North American sills and lavas. Transfer of residual liquid from the compacting lower solidification front to the dilating upper solidification front resulted in characteristic chemical

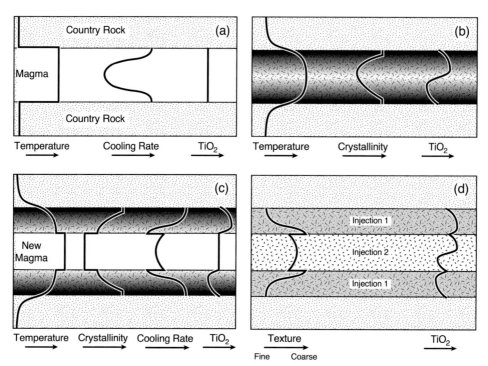

Figure 4.4 Sequential depiction of the formation of Beacon-type sills, illustrating the ongoing effects of reinjection events on the ensuing textural and chemical variations. (a) Initial injection of thermally and chemically homogeneous magma. (b) After a period of cooling, the magma partially crystallizes, and compaction redistributes the interstitial liquid, producing a sigmoidal variation in $TiO_2$ (and other incompatible elements). (c) A new injection of magma breaks apart the partially crystallized sill at its weakest point (roughly in the center) and displacing the previously formed rock to the outside. The newly injected magma cools most rapidly at its margins, whereas the surrounding partially solidified magma now being reheated cools more slowly than before. (d) After final solidification, the sill preserves the textural and chemical discontinuities marking the reinjection horizons (after Zieg & Marsh, 2012).

and mineralogical effects, such as the depletion of the lower half of the sill and the enrichment of the upper half of the sill in incompatible elements (e.g., $TiO_2$, Zr) and modal granophyre. Based on thermodynamic models, most of the compaction occurred at a crystallinity of roughly 33 percent. Geochemical profiles are distinctly segmented, suggesting that the sill was repeatedly split and reinjected with fresh magma near the center after compaction had redistributed the interstitial liquid within the partially solidified magma. The recognition of evidence for reinjection and compaction in a macroscopically uniform, relatively quickly cooled sill suggests that these processes may be common in the construction of sills and other mesoscale igneous intrusions in general. And, as will become clear in Chapter 5, in the detailed MgO profiles of the Basement Sill, and in the isotopic profiles through the Peneplain Sill and the Dais Intrusion (see Chapter 10), this style or repetitive injection is evidently a commonplace feature of dolerite sill complexes. A broadly similar, but even more detailed analysis of the 250 m thick Black Sturgeon Sill, Nipigon, Ontario shows a similar history of multiple injections (Zieg & Wallrich, 2018). And Archean dolerite sills may well have undergone similar histories of emplacement and in situ differentiation, giving rise to conditions conducive to gold mineralization (Hayman et al., 2021). A generic depiction of this process of reinjection is given by Figure 4.4 and is presented as a standard state of sill development that is most often not at all obvious. Telltale signs of this or similar processes will show up from place to place in the textures and chemical compositions of the Ferrar sills.

The major lesson learned from studies of this type is that the larger the mass of the intrusive, the more individual injections are expected to have formed it.

## 4.3 System Development

As premature as it may seem at this point to discuss the overall detailed development of the magmatic system, both the understanding and development of the ongoing presentation are greatly facilitated if this is done here. This is a major conundrum.

As discussed, from a consideration of the field relations, basic hydraulics, and chemistry of the sills, dikes, and lavas, this Ferrar system most likely developed from the top down. It would be convenient and reasonable to state further that the system began as early dike swarms pervading many parts of the area vented to produce local piles of tephra and lava, capping the eruptions, and forcing smaller dikes to coalesce into more substantial feeders that eventually evolved into producing a stack of sills. But, as is evident everywhere in this region, there is little sign of major rifting and an extreme paucity of any dikes, let alone any dike swarms, able to transport, erupt, and establish magmatic centers on the scale found in Bull Pass. This is the conundrum having no obvious simple explanation that does not defy the field relations, although the possibilities will become apparent as the presentation proceeds, and this will be returned to in much more detail in Chapters 12 and 13.

For the time being what seems clear is that local vents were established in the form of phreato-magmatic cauldrons, associated with the earliest sills breaching the surface, from which emanated voluminous tephra (Mawson Formation) followed by thick welts of basaltic lavas (Kirkpatrick Basalts). This lava increasingly capped the subaerial system, forcing ensuing rising magma to further expand the high-level associated sills and perhaps nucleate more vents, which is well recorded at the highest levels at places like Shapeless Mtn., where there is almost every imaginable transition between dike, sill, and tephra, with major blocks of Beacon fragmented, tilted, and encased by dolerite (Korsch, 1984). These occurrences suggest a close succession of emplacement pulses, perhaps reflecting episodes of volcanic eruptions, as is also seen in the sills themselves.

The capping effects of an increasing thickness of volcanic overburden and the emplacement of the uppermost sills forced the deeper system to further concentrate into a centralized feeder. This is also encouraged by the characteristic period of major volcanic episodes, which may locally be quiescent for years to 10s of years to even 100 years or so. These scales of down or repose time translate into the full solidification of dikes and fissures as thick as about 50 m. Through solidification at the smaller scales the system became sealed, allowing with renewed activity for it to become over-pressured enough to emplace the massive lower sills. The two massive lower sills are each sequentially denser than the next overlying sill and show, as already mentioned, distinct contact relations of having been formed in relatively rapid (thermally) succession. The final result is a central feeder or stalk zone from which emanates sills that climb and jump locally to form, in essence, a magmatic structure in the form of a fir tree (see Figures 1.1 and 1.2). This basic structural form has also sometimes been previously implied, but it is particularly clear here on many levels (e.g., see description of Basement Sill in Chapters 5 and 6).

The overall developmental sequence is one of increasing localization with time of crustal conduits and vents from regional swarms of thin dikes (1–5 m), with short thermal relaxation or freezing times, to massive central feeders that sustained the system over long times ($\sim 10^4$ to $10^5$ years, or more), giving rise to areally extensive sills of increasing volume and bulk density downward in the system. There is also good reason to believe, as will be discussed in Chapter 6 in more detail, that this basic intrusive structure may continue downward to much deeper levels in the crust. And it is important to appreciate that the system did not necessarily develop due to density or buoyancy, although this certainly had an effect, but from the very beginning by becoming over-pressured and forcing dense magma to traverse the crust and erupt to the surface.

# 5

# Nature and Distribution of Individual Sills

## 5.1 Introduction

The four prominent sills (Mt. Fleming, Asgard, Peneplain, and Basement) were studied in detail by mapping their areal distribution and by detailed stratigraphic sampling at intervals of generally either 15.25 or 30.5 m (i.e., 50 or 100 feet) from the upper to the lower contact (Figure 3.1). (In some instances, as in the Beacon Sill study, sampling was on a much more detailed level, which will be mentioned when necessary.) The sampling profiles encompass most of this region (Figure 5.1). For obvious reasons, the most complete sampling profiles, by far, were made of the Basement Sill (18), compared to the Peneplain Sill (8) or the Asgard (4) and Mt. Fleming (2) sills. A good deal of additional cursory or grab sampling was also done throughout the region on all the sills as a matter of education and convenience. And much time was spent in generally studying the rocks themselves in place. So, the impressions imparted herein rest on a great deal more than these specific sampling profiles; contacts and interiors were often walked out over large regions, sometimes repeatedly. The detailed sampling is a direct reflection of the availability of good sections for sampling (i.e., safe areas with well exposed upper and lower contacts) and the apparent petrologic value of the rock. Outcrop decreases dramatically with elevation, such that relatively scant information was collected on the upper two sills. But the petrologic heart and richness of the system is in the Peneplain and Basement sills, each of which exhibits a vast assortment of outcrop. As a beginning, characteristic magnesia profiles are useful to typify each sill (Figure 5.2), which is taken as a proxy for Opx (and lesser Cpx) phenocryst concentration.

## 5.2 Mt. Fleming Sill

Occurring at the highest elevations against the Polar Plateau, this sill is relatively thin (~100 m), black, and in outcrop often rusty with distinctive (vesicle-like) circular wind abraded erosion pits. It is unclear whether the tendency of this rock

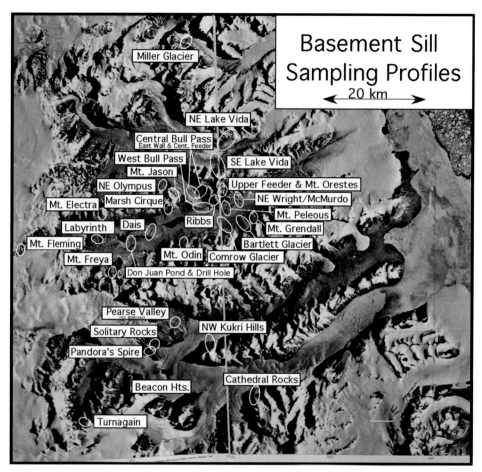

Figure 5.1 The locations of the principal detailed sampling profiles through the Basement, Peneplain, Asgard, and Mt. Fleming Sills

to form these pits is due to incipient vesiculation, which is not obvious on fresh handsample faces, or simply due to its erosional exposure via intense sand blasting at this elevation. It is perhaps the most structurally complex sill in that it is disjointed and splayed with common local dike-like offshoots, perhaps reflecting the near proximity of the original free surface. At Mt. Fleming, Shapeless Mt., and Allan Hills, it (or its equivalent) has also spawned other small sills and dikes that intersect the original surface and turn into all combinations of mixtures of tephra and dolerite. The sampled profile at Mt. Fleming is texturally and chemically quite uniform with MgO at $4 \pm 0.5$ (wt.%).

## 5.3 Asgard Sill

This is the most distinctive sill at higher levels (~5,500 ft.) in the Asgard Range, forming distinctive caps on buttes of Beacon sandstone and shear walls in the

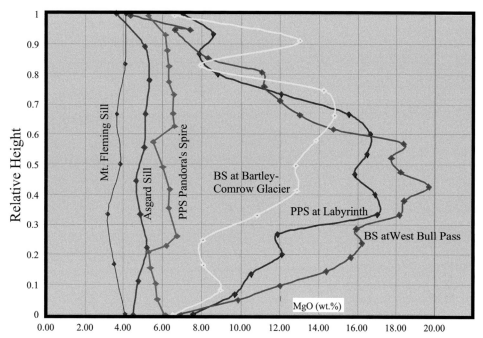

Figure 5.2 Representative chemical stratigraphy (MgO wt.%) through the Basement Sill (BS), Peneplain Sill (PPS), Asgard Sill, and Mt. Fleming Sill at specific locations. There is a systematic increase in MgO downward in the section, and there are strong differences in the Peneplain Sill between Pandora's Spire at Solitary Rocks and the Labyrinth, where the Opx Tongue exists, containing nearly as much Opx as in the Basement Sill at West Bull Pass. Note also that the Upper Lobe of the Basement Sill in the Asgard Range at the Bartley-Comrow Glaciers area, across from Bull Pass, is also rich in Opx.

Inland Forts, which R. F. Scott visited and used for navigation when on the Polar Plateau in 1903. At 200–250 m thick, it is apparently absent north of Wright Valley, as it does not appear in the buttes of the upper Olympus Range. To the south, it is well exposed in Beacon valley, where it has been called the Beacon Sill and has been studied in some detail by Zieg and Marsh (2012; see also McKelvey & Webb, 1959). Although sometimes splayed looking faulted with good thermal jointing (Figure 5.3), it is more commonly continuous and structurally simple. It is black with a slight reddish hue in places and uniform with a medium grained texture, aside from chilled margins. The texture does, however, show a subtle spatial unevenness in the central regions, perhaps suggesting repeated remobilization or reinjection during solidification. The sampled profiles, at Mt. Freya, Beacon Heights, and Mt. Electra, are almost uniform throughout in MgO at 5 ± 0.5 (wt.%). The highly detailed sampling at Beacon Heights shows systematic, symmetrical variations in major and trace elements that reflect a systematic process of multiple reinjections leading to the final formation of the sill (Zieg & Marsh, 2012).

Figure 5.3 A step change in the Asgard Sill in Upper Wright Valley with distinctive columnar jointing, perhaps indicating emplacement from east to west. Also note the style of rib-like erosional forms of the columnar jointing in the upper left, which is broadly similar to the Ribs Section in Bull Pass.

## 5.4 Peneplain Sill

Always appearing just above the Kukri erosion surface and extending at least from the south side of Victoria Valley in the north to Finger Mountain and Turnabout Valley in the south, this sill (~350 m thick) is distinctive in its massiveness and uniformity in form and, in most places, texture. This is the picturesque sill in the north wall of Finger Mountain (Figure 2.13), often-photographed, beginning with Albert Armitage of R. F. Scott's *Discovery* Expedition in 1902; Hamilton (1965) shows particularly striking pictures. It does not, however, extend northerly beyond the Olympus Range. Although appearing at Mt. Jason and near Mt. Hercules, just west of Bull Pass, and to the east above Bull Pass, it does not appear in the southwest wall of McKelvey Valley or in the high buttes of the western Olympus Range. It exhibits a wide range in chemical variation reflecting its gross appearance. At Solitary Rocks in the Taylor Glacier, it is uniform at ~6 wt.% MgO and is black, dense, and of medium grain size throughout. Here and there are occasional slight grain size variations in the central areas, indicating multiple or pulsatile injections like that displayed by the Beacon Sill (Zieg & Marsh, 2012). Twenty-five kilometers north, in Wright Valley, it contains a Tongue of Opx and up to 17 wt.% MgO (Labyrinth, west Wright Valley) and up to 15 wt.% MgO in the Olympus Range at Marsh Cirque and in the first valley immediately west of Bull Pass, leading into McKelvey Valley. Where it does contain the Opx Tongue it is more reddish, coarser grained, and with occasional distinct curvate and indistinct "inch-scale" modal layering as at, respectively, Marsh Cirque and the Labyrinth. It weathers into plates, slabs, and cliffs.

There is, nevertheless, a chance that this is not the same sill in Taylor and Wright Valleys; there might be two separate sills at this horizon. The absence of an Opx Tongue and the slightly higher elevation of the sill in the Taylor Valley region versus that in the Labyrinth might suggest two sills. Yet, the striking continuity of the sill along the south wall of Wright Valley and its apparent continuity southward, albeit disjointed and partially snow covered, across the central and eastern Asgard Range to Taylor Valley argues for one continuous sill. And the Basement Sill horizon is also lower at the west end of Wright Valley than it is elsewhere. Apparently the Opx Tongue, which is less developed in the Peneplain Sill in Wright Valley, is also much less spatially extensive than it is in the underlying Basement Sill. This may also be partly due to sampling access, which, because it does not extend more northerly into and across McKelvey Valley, in the Olympus Range the sampled areas must be on its northern margin where any Opx might have been segregated away. The apparent volume of the Peneplain Sill is about 650 km$^3$.

## 5.5  Basement Sill

With a volume of about 1,050 km$^3$ and underlying the entire region, this ultramafic sill is the core of the Ferrar system in the Dry Valleys. At the time of its emplacement a vast region of perhaps 10,000 km$^2$ of the crust was literally floating on this sheet of magma. It was emplaced in the granitic country rock just below the Kukri erosion surface; often a matter of only a few tens of meters below, it precisely follows this remarkably regular surface. The ability to attain this careful attitude certainly reflects the ability of the overlying crust to remain intact and also the formation of expansion fractures formed in response to the erosional unearthing attending formation of the Kukri Peneplain. It extends at least from Cathedral Rocks on the Ferrar Glacier, and possibly much further south, to nearly 100 km north to Wheeler Valley on the Miller Glacier; it does not appear to extend further north into Killer Ridge, but it may extend northwestward. On the west, a piece of its upper contact is exposed near the Labyrinth in Wright Valley on the north wall of the Dais and its eastern limit is about 7 km east of Bull Pass. It is distinctive in appearance, relative to the overlying Peneplain Sill, with a slight olive-brown hue, being medium-to-coarse grained, and weathering into a lumpy

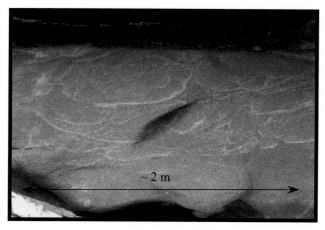

Figure 5.4 Characteristic typical plagioclase stringers or schlieren found through-out the Opx Tongue rock of the Basement Sill; indicative of the direction and nature of the shear flow of emplacement of the Opx-rich granular slurry

and often intricately, but also smooth rather than sharply ragged, sandblasted carved landscape.[1] This all reflects the pervasive presence of the massive Opx Tongue, which extends virtually to every corner of the sill known so far.

A diagnostic field indicator of the Opx Tongue presence is the occurrence of thin wispy stringers or schlieren of fine-grained plagioclase that have formed in response to shearing of the concentrated Opx granular material (Figure 5.4). Shearing creates local dilatant zones into which sifts small plagioclase grains. These stringers are commonly 0.5–1 m in length and 0.5–1 cm in thickness, undulate, and generally approximately coplanar with the form of the sill itself. In some places these are well-formed anorthositic layers (1–2 cm thick) continuous for 1–3 m and sometimes also small, isolated stock-like vertical fingers, 1–3 cm in diameter and 5–15 cm tall, that have evidently formed in response to the final settling, compaction, and pocket slumping of the initially fluidized magmatic mush.

In an effort to reproduce these features experimentally, the extensive Opx dune sands at southwest Bull Pass were made to avalanche along the present angle of repose. Troughs were also made with Plexiglas walls in order to visualize the schlieren in the true approximately horizontal form of a shear flow. All these experiments, regardless of the setup, always produced schlieren; there was, in fact, no way to prevent them from forming with any movement of the sands (Figure 5.5). In all instances they were accurate indicators of the nature of the overall motion of the granular medium.

---

[1] When R. F. Scott referred to the sills in the Upper Taylor Glacier region as being "rich brown in color" it is the Basement Sill he is talking about, presumably from his close traverse at Solitary Rocks.

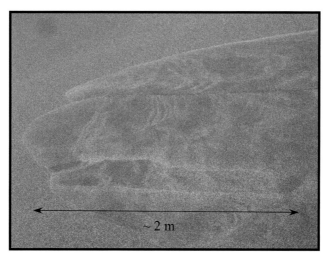

Figure 5.5 The experimental formation of plagioclase stringers by stimulating avalanches in the Opx-rich dune sands at West Bull Pass. Here the Opx Tongue is extensively wind eroded into substantial dunes of coarse pyroxene and much smaller plagioclase; any disturbance promotes petal-like shearing avalanches, causing segregation of the small plagioclase grains into thin beds at the base and edges of the avalanches.

## 5.6 MgO Profiles

Eighteen analyzed stratigraphic profiles characterize the spatial variations in composition in sufficient detail to gain insight into the mode of emplacement of the Basement Sill, some of which are given in Figure 5.6. Each profile is marked by phenocryst-free chilled margins with a common, consistent MgO content of about 7 wt.%. The highest concentration of MgO is generally in the central part of the sill; profiles with the highest MgO content (~20 wt.%) are the most symmetrical. For future reference, to give some idea of the tradeoff between Opx modal abundance and whole rock MgO content: For ~55 vol.% Opx phenocrysts (i.e., solids at maximum packing) with a composition of $En_{80}$ (see Chapters 7 and 8) mixed with melt of a background composition of 7 wt.% MgO gives a whole rock MgO content of about 20 wt.%. And, as will be shown in Chapter 8, the Opx crystals are generally larger and more En rich toward the sill interior.

The maximum concentration of sill MgO content decreases outwards from the general area of west Bull Pass. At west Bull Pass the sill is gorged with phenocrysts and the MgO content has a parabolic distribution, reaching a maximum of about 20 wt.%; this is also found nearby to the west at the Dais and southeast Lake Vida profiles, although at the Dais approximately half of the section is buried. Nowhere else are the profiles quite as fully parabolic. Strong steps and deep recessions in MgO content are most likely related to the episodic

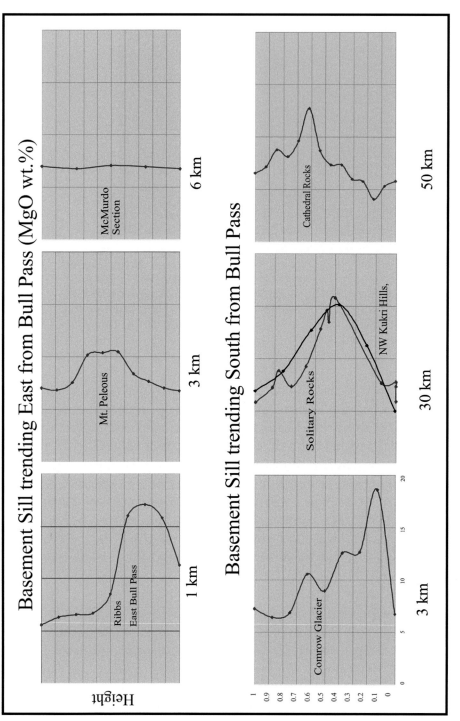

Figure 5.6 Stratigraphic compositional profiles (MgO wt.%) through the Basement Sill outward from Bull Pass; each profile is normalized to a height or sill thickness of unity. (a) The upper series goes eastward from Bull Pass, toward McMurdo Sound along the north wall of Wright Valley, and the lower series goes southward from Bull Pass across to the Comrow Glacier (upper Lobe of BS) and ending at Cathedral Rocks. (b)The upper series goes north from Bull Pass through Victoria Valley and ending at the Miller Glacier. The lower series goes west from Bull Pass to the Dais Intrusion and Don Juan Pond, which includes samples from the drill core there. (Here the actual thickness of each section is also given.)

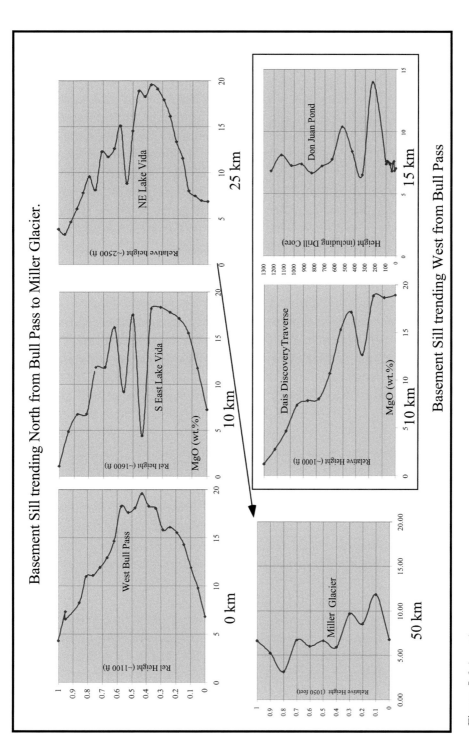

Figure 5.6 (*cont.*)

nature of emplacement, which will be discussed in some detail later (e.g., Charrier & Marsh, 2004, 2005, 2021; Charrier, 2010; Zieg & Marsh, 2012). Aside from the steps and recessions, there are two major spatial characteristics of these profiles: the amplitude and position within the sill of the elevated MgO concentration. The MgO content is weaker and higher in the sill with increasing distance from Bull Pass. These characteristics are also specific to certain sectors of the sill, namely, north, south, east, and west, which most likely reflect the stress field associated with rifting and sill emplacement. McMurdo Sound ocean depth strikes almost north–south. Perhaps similar, the strongest development in concentration is north from Bull Pass, as shown by the southeast and northeast Lake Vida profiles. Twenty-five kilometers further north, however, at Wheeler Valley (about 50 km from Bull Pass) the maximum concentration is weak, at about 12 wt.%, and it is uncharacteristically near the sill base.

South from Bull Pass the next nearest profile is immediately across Wright Valley at Comrow Glacier; this is the continuation of the upper lobe or duplex of the Basement Sill. Here the maximum MgO content is about 18 wt.% and is narrowly distributed near the base of the sill. With increasing distance, the profiles at northwest Kukri Hills and Solitary Rocks are similar, with a maximum of about 15 wt.% MgO and the concentration is about in the center of the sill; these locations are about equidistant from Bull Pass. At Cathedral Rocks, the most distant southern profile, about 50 km from Bull Pass, the maximum MgO content has decreased to about 13 wt.% and is narrowly distributed in the central part of the sill.

East of Bull Pass, over a distance of 7 km, the Basement Sill dramatically decreases in thickness to a horsetail splay of thin (10 cm) wispy dikes extending several hundred meters beyond the sill tip proper. One of these wispy dikes curves upward, cutting the upper lobe of the overlying Basement Sill and also the Peneplain Sill. This indicates a late event perhaps reflecting the cessation of overpressure associated with the overall system dynamics during final collapse or relaxation of the system, forcing crystal-free, low viscosity melt further outward in the system. The Opx Tongue also shrinks and ends in a blunt tip after about 5 km from Bull Pass. The leading tip of the sill for about 2 km beyond the tongue thus contains no phenocrysts and is of uniform composition at about 7 wt.% MgO. The Opx Tongue about 2 km away has a maximum MgO content of about 10 wt.% in the center of the sill. Still nearer Bull Pass, the East Bull Pass (or Ribs) profile has a maximum MgO content of about 17 wt.% concentrated in the lower part of the sill (Figure 5.7).

To the south and west the sill migrates lower into Wright Valley forming the Dais Intrusion (see Chapter 7) at the east end of the Dais topographic feature. On the north side of Wright Valley, the sill declines smoothly in elevation from west

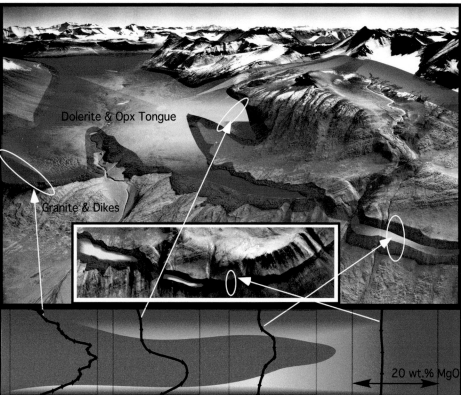

Figure 5.7 The exposure and distribution of the Basement Sill in Bull Pass, looking north, and the northeast wall of Wright Valley (US Navy aerial photo). The upper photo shows the how the sill fills Bull Pass from the very thick section on the west

Bull Pass to the Dais area, maintaining its thickness or possibly slightly thickening. On the other side of the valley (south wall of Wright Valley), the sill appears as a series of large, southward propagating coalescing lobes. The lower contact of each lobe often climbs along a Devonian dike with the upper contact following until encountering the overlying Peneplain Sill, against which it quenches (see Figure 5.8). This pattern of climbing and quenching continues westward to the area of Mt. Odin where the sill dramatically climbs as if to form an eruptive vent but is apparently capped by the Peneplain Sill (see Chapter 6). At the base of this feature the sill is massive and shows pervasive layering laterally over hundreds of meters between massive brows of pyroxene units and fine-grained anorthosites, the nature of which will later be described in more detail. The sill continues westerly, dipping strongly (~60°W) toward the Dais in the center of Wright Valley, directly southwest of Lake Vanda. Further west along the south wall of Wright Valley, there is yet another lobe of the Basement Sill at Don Juan Pond. Here the sill is at a much lower elevation and the lower contact is buried beneath the valley floor fill. Subsurface samples were obtained from the nearby drill hole associated with the McMurdo Dry Valleys Drilling Project (e.g., Mudrey et al., 1975; McGinnis, 1981), which did not reach the lower contact. The higher concentration of MgO at Don Juan Pond is in the lower part of the sill, reaching a maximum of about 14 wt. %, and is double peaked. At the east end of the Dais the Opx-rich magma ponded to form a small, but magnificent in many ways, layered intrusion (Dais Intrusion), which will be described in detail in Chapter 7.

The overall form of the Basement Sill thus appears as a series of lobes and approximately horizontal massive blades propagating outward from a common central feeder region in Bull Pass (see Chapter 6). The southward propagating lobes appear to coalesce in the fashion of the leading lobes at the front of massive lava flows, characterized, in essence, as a highly viscous fluid moving within an elastic bag or envelope. The northerly and northwest quadrants propagate into Victoria and McKelvey Valleys as an expanding single large blade or horizontal fin, extending as far north at the Miller Glacier and Killer Ridge. The leading edge of the sill is exquisitely exposed over a vast area in Victoria Valley where it is generally an intricate assemblage of small (10–20 cm) to tiny (1–5 cm) dikes and

---

Figure 5.7 (*cont.*) side (LHS front) and greatly thinning in propagating to the east toward McMurdo Sound. The presence of the Opx Tongue in the sill center is clear on the east due to its sensitivity to wind erosion. In the lower photo the position of the Basement Sill is indicated in color and the position of the Upper Lobe is also schematically emphasized. The lower-most diagram specifically shows the variation in MgO (wt.%) at various spatial positions, beginning at West Bull Pass and extending eastward toward McMurdo Sound where the presence of the Opx Tongue becomes progressively smaller.

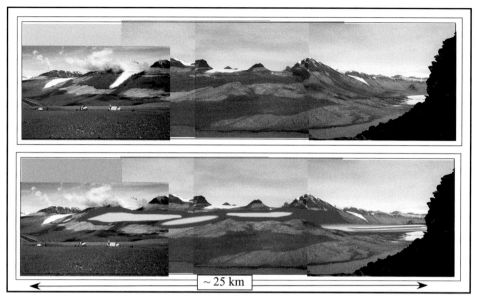

~ 25 km

Figure 5.8 A panoramic view showing the form and distribution of the Basement Sill in the South wall of Wright valley across from Bull Pass. The upper photo is enhanced below to clearly depict the lobate nature of the sill here. (Upper) The Peneplain Sill forms the mountain peaks of the Asgard Range and the lower lobate sill-like features in the valley wall are of the Basement Sill. The step-like nature of the lower contact in the center reflects the tendency of the magma to follow along boundaries provided by the regional (much older) dykes. The lobes on the left are those stemming from the Upper Lobe of the Basement Sill propagating outward from Bull Bass. And on the extreme west (RHS) the sill dips down, ponding, and forming extensive layering leading to the Dais Intrusion. The overall Basement Sill structure indicates that the sill propagated southward as a series of laterally expanding and coalescing lobes. (The dark, sill-like, mass below the Basement Sill is scree.)

sills. On a larger scale, the stacked nature or lobe-like nature of emplacement of the complex is exemplified in the detailed field relations found in Bull Pass (Figure 5.9), which may be a microcosm of the mush column nature of the entire Ferrar complex; much more detail given next.

This characteristic of the system to form coalescing lobes representing repeated emplacements is exemplified in the MgO profiles exhibited by the overlying Peneplain Sill in going outward from Bull Pass (Figure 5.10). In spatial relations these profiles, beginning at west-central Bull Pass (NW Bull Pass profile), go along the Olympus Range, to Mt. Jason. Marsh Cirque, Labyrinth, and then all the way over to Pandora's Spire at Solitary Rocks, which is far from the Opx Tongue. These profiles through the Olympus Range each show strong concentrations of MgO, reflecting the presence of the OPX Tongue, and also show all the

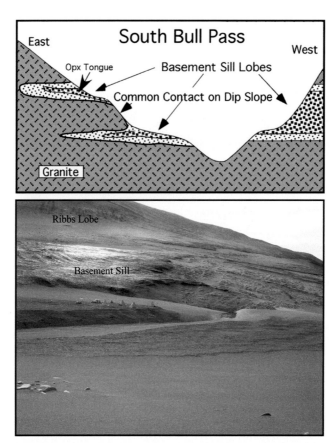

Figure 5.9 The field relations in south Bull Pass. (upper) A schematic cross-section illustrating the lobate nature of the Basement Sill on the east side of south Bull Pass. The sill delicately coats the dip slope between the upper Ribbs section and the lower lobe that propagates easterly toward McMurdo Sound. (lower) The wind-polished chilled margin of the Basement Sill on the dip slope above camp. On the upper right the Basement Sill trends easterly, and the strip of sand just below marks the presence of the Opx Tongue.

characteristic features (i.e., plagioclase stringers, anorthosite segregations, etc.). The strongest concentration of MgO, at 18–20 wt.%, however, is at Marsh Cirque, which is west of Mt. Jason and just south of Mt. Hercules, whereas Mt. Jason and NW Bull Pass, both closer to Bull Pass, have lower concentrations. This may suggest that the style of feeding the Peneplain Sill may have been from more than one location. The Marsh Cirque sector may have been fed by the strong lobe of the Basement Sill in southwest Bull Pass, which has coarse OPX Tongue material extending in a climbing contact through to the upper contact. The NW Bull Pass, Mt. Jason, and the Labyrinth sections, instead, may have been fed from central Bull Pass, which will be discussed much more fully in Chapter 6. Cursory

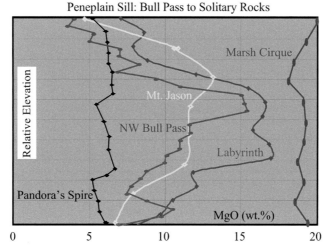

Figure 5.10 The variations in MgO (wt.%) in the Peneplain Sill from above the west wall of Bull Pass to Mt. Jason, Marsh Cirque, the Labyrinth, and across to Pandora's Spire at Solitary Rocks. There is a progressive decrease outward from Bull Pass, and the strong concentration at Marsh Cirque may suggest multiple filling locations.

examinations of the OPX crystal sizes indicates a steady diminishing in size from NW Bull Pass to the Labyrinth.

In essence, from whatever perspective these sills are examined, basic field relations, crystal textures, or chemical compositions, it seems that all roads of inquiry lead to Bull Pass.

# 6

# Bull Pass Geology

## 6.1 Introduction

The intricate, deceptive, and stunningly preserved field relations make Bull Pass perhaps the most telling and fascinating region in all the Dry Valleys. This is because, fortuitously or not, the elevation of Bull Pass is exactly at the regional base of the main Basement Sill. The location of the main feeder or vent zone is here, the style of emplacement of the whole system is exemplified here, and the overall complex is quite likely in the very throat of an overlying caldera complex. This is all coupled with the clarity in aerial photos of the Opx Tongue, which, due to its erosiveness, serves as a tracer of the general flow dynamics of emplacement. A highly useful summary of the larger scale field relations is given by Turnbull et al. (1994).

## 6.2 Detailed Field Relations

### 6.2.1 South Bull Pass

As mentioned in Chapter 5, from the south lip of Bull Pass to the immediate west the sill is a single massive unit, and to the east the sill dramatically thins and dies out over a distance of about 7 km. A gully eroded into this east limb in the north wall of Wright Valley shows that the sill may also thin northeastward into the hillside (Figure 5.7). In the east wall of Bull Pass, just above the thinning east lobe, but below the upper duplex lobe, is another lobe of the sill. It exhibits massive columnar jointing and is called the Ribs Lobe. The upper chilled margin of the east lobe and the lower chilled margin of the Ribs Lobe show the dolerite to be quenched on the *surface* of the dip slope, which means that these were co-propagating laccolith-like lobes of the Basement Sill (Figure 5.9). As the major sill propagated southeastward from central Bull Pass it split into contiguous upper and lower lobes. Careful examination shows here and there small, delicate bits and

Figure 6.1 Geologic field relations of the Basement Sill in the southeast wall of Bull Pass. The distinctive rib-like pattern of incipient columnar jointing marks the overall lobe-like nature of the emplacement process, and upper extension (Feeder Zone) is the beginning of the Upper Lobe that propagates southerly through the Asgard Range and reconnects at the Kukri Hills with the lower, more westerly lobes, that go on to Solitary Rocks (US Navy aerial photo).

pieces of chilled margin against the granite dip slope country rock over large areas on the slopes east of south Bull Pass. Interesting enough, the much older dike swarm is an excellent strain marker for the emplacement of the Ribs Lobe, and on its southerly end it seems to show little deformation as if the space occupied by the dolerite was mined or stoped out and not simply forcibly pushed apart; this may well reveal a critical geometric indication of the overall emplacement mode.

The lower contact of the Ribs Lobe extends northerly along this wall into central Bull Pass where it wraps back into the east lobe (Figure 6.1); the actual connection is obscured by cover. The upper contact of the Ribs, which is knife sharp against Devonian dikes and plutons, continues northerly but sweeps up section to form an apparent connection (informally called the "Feeder Zone") with the overlying upper duplex lobe just beneath the Peneplain Sill, continuing southward across to the Comrow Glacier area and then continuing southerly to the Kukri Hills, as described in Chapter 5. The sill here climbs almost vertical for about 600 m to the horizon of the Peneplain Sill (Figure 6.2). The opposite (north) contact of the Feeder Zone is easily traceable downward to the upper contact of the Basement Sill

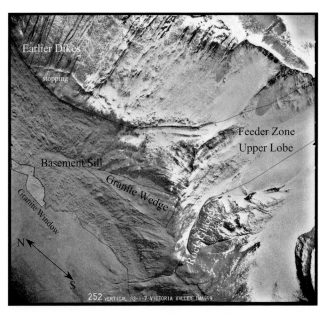

Figure 6.2 Upper and lower contacts of the Basement Sill in the east wall of Bull Pass; the sill climbs and splits here to form the Upper Lobe via the Feeder Zone. The large granite block riddled with old dikes, on the left, is sinking into the sill and a wedge-like segment marks the incipient stages of stoping. The overall lumpy and jointed nature of the sill marks the presence of the Opx Tongue material (US Navy aerial photo).

in central Bull Pass where it continues northward into Victoria Valley at its usual elevation (Figures 6.2 and 6.3).

In rising to form the Feeder Zone, the sill itself forms the sill duplex separated by a large (~100 × 350 m) wedge of granite (Figure 6.2). Where the two units meet at the foot of this wedge there is a decrease in grain size, but no chilled margin. The lower sill of the duplex goes to form the Ribs laccolithic lobe and the upper sill forms the Feeder Zone, connecting to the upper lobe beneath the Peneplain Sill. The apparent stoped wedge of granite country rock also wedges laterally. That is, the upper contact is cliff like, indicating a thickness of at least 10 m, but the lower contact thins dramatically into a thin sheet (~1–3 cm) that is pervasively and delicately brecciated by the dolerite. That the dolerite is hardly chilled indicates the rock of the granite wedge was already hot during brecciation. Judging from the orientation of the much older original dikes within the wedge relative to those in the nearby similar country rock, the wedge did not become stoped and fully dislodged, but is apparently rooted into the country rock at depth normal to the hillslope. This may also be further evidence that the sills are themselves also wedging out to the northeast in the propagation direction, as depicted in Figure 5.9. Above the upper dolerite enveloping the granite wedge, a

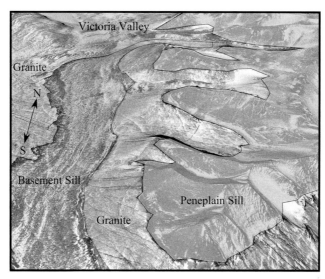

Figure 6.3 Basement Sill and Peneplain Sill in southeast Victoria Valley and extending easterly into Victoria Valley. The ragged nature of the upper and lower contacts of the Basement Sill, as opposed to the tight contacts of the Peneplain Sill, reflect its aggressive nature of emplacement, extensively melting the country and stoping its way upward. The lumpy and apparent layered fabric of the Basement Sill indicates incipient processes of layering in the Opx Tongue rock (US Navy aerial photo).

thin (50 cm) wedge-shaped dike extends from the dolerite some 50 m upward into the overlying granite, apparently quenched in the act of stoping a sheet of granite; a part of this can be seen in the photo of the climbing contacts (Figure 3.2).

Following the Basement Sill northeast into Victoria Valley, the entire width of the exceedingly well-exposed sill is delicately spread across the gentle granite slopes of the south wall of the valley (Figure 6.3). The sill continues easterly and terminates in a broad blunt tip, unlike the eastern lobe at Bull Pass, and turns back, climbing strongly in elevation to the southwest, propagating beneath Mt. Orestes and apparently meeting (the exact relations are snow covered) the Feeder Zone found in east central Bull Pass. This suggests that this lobe is not connected to the terminus of the nearby east lobe of the Basement Sill about 5 km to the south in the north wall of Wright Valley.

The kinematics of emplacement of the sill blunt tip is exceptionally well exposed here in north Victoria Valley. In places the granitic country rock has a distinctive fine gneissic banding, clearly revealing this wall rock to have been broken into a continuum of meter size domino-like parallelepipeds that have been jostled about to make room for the sill. The sill emplacement process to the west into McKelvey Valley is distinct from this in that over long distances (kms) the propagating contact is a series of aggressive, often ragged, small (10s of cm), knife-sharp lobes cutting into the country rock. The style of advancement and filling is reminiscent of the pulsative inflationary process for emplacement of large pahoehoe flood basalts (Self et al., 1996, 2000; Anderson et al., 1999). In some places these features are broadly similar to the stepped wedges and fingers depicted for the growth of the sills in the North Rockall Trough by Thomson and Hutton (2004), which will be discussed more fully in Chapters 12 and 13. Nevertheless, this entire intimately exposed contact throughout this region is a superior record of the mechanics of sill propagation.

The Feeder Zone near Mt. Orestes that goes to form the Upper Lobe, above the Ribs-Granite Wedge area, was sampled along the arête between Mt. Orestes and the next peak to the east. The maximum MgO content is 9 wt.% near the north margin and then levels out to 5 wt.% for the remainder of the profile (~300 m sampled every 30 m). But because the geometry of this feature is not clear, this profile may run at a highly acute angle to the contact. Nevertheless, the MgO content of most of this material is markedly less than that of any part of the Basement Sill, from which it emanates, perhaps showing the loss of Opx phenocrysts as the dolerite ascends.

### *6.2.2  Central Bull Pass*

To the west, across the valley of central Bull Pass at the foot of the Olympus Range, the upper contact of the western, massive, lobe is scree-covered, but a distinct knick point in the landscape, marking the change in slope, coincides with the onset of dolerite float at the expected location of the contact. And a detailed satellite photo also seems to show this contact (Figure 6.4). The lowermost portion forms a massive intrusive zone in the floor of central Bull Pass and even more intricate field relations are found in the east wall, which will be described.

Along the floor of central Bull Pass proper, beginning at the south lip of the Pass, the sill can be followed northerly in the incised drainage, but the exact relations between the west and east sectors of the Basement Sill become intricately confused. That is, however carefully the contact relations are traced into this area just south of the small frozen lakes (i.e., from the duplex–Ribs–eastern lobe areas or the western massive sill lobe), the inescapable impression is that the textures coarsen and the sill becomes massive. Areas of clear granitic country rock to the east followed into this central zone become completely recrystallized and in places blend imperceptibly into dolerite (see Granite Window in Figure 6.1). Taken altogether, this suggests that the sill here either becomes unusually massive, perhaps forming another lobe in the subsurface, or that this is the top of a deeper feeder zone. That this may indeed be the central feeder location is further suggested by the near vertical dip of the associated plagioclase stringers in the

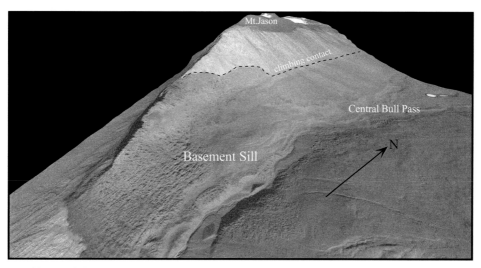

Figure 6.4 Satellite image of Bull Pass, showing possible obscure contact of the Basement Sill (center left) in the west wall of the valley. Notice also the lumpy, coarse character of the Basement Sill in the immediate foreground and continuing upwards forming a distinct climbing contact (complements of Quickbird Imaging, 2008).

basal section of west Bull Pass near the field of sand dunes. These markers become increasingly steep with the approach to central Bull Pass, culminating with fully vertical cylindrical bundles of stringers (see Figure 6.5). This is a common sorting process in granular materials, known as shear localization, as mentioned with Figures 5.4 and 5.5, which is highly similar to the patterns formed by heterogeneous agricultural grain in being drained from silos (e.g., Wojcik & Tejchman, 2009).

In the east valley wall, just north of the confluence with Orestes Valley (see Figure 6.12) the upper contact of the Basement Sill again splits into a duplex style, forming a series of benches in the granite slope (see Figure 6.6). This marks the penetration into the granite of a strong wedge of dolerite about 50 m above the sill proper, perhaps in response to stoping, much as in south Bull Pass (Figure 5.9) and also, from the pattern of the upper contact, suggests the dolerite may have traveled extensively upslope, perhaps to feed the overlying Peneplain Sill; cursory examination of the Peneplain Sill here shows no obvious signs of this, but it may have been more westerly. The granite country rock between the sill proper and this upper wedge is extensively melted to the point that it, the granite, has entirely lost its original texture and in many areas blends seamlessly into highly contaminated dolerite (Figure 6.7). Similar areas of extensive melting of the granite are found southward all along this upper border of the dolerite. And turning northward into Victoria Valley the same upper margin of the Basement Sill appears diffuse,

Figure 6.5 Bundles of vertically plunging circular plagioclase stringers or schlieren in central Bull Pass, indicating the vertical nature of the magmatic flow direction (see ski stick for scale)

representing a major "climbing contact" along with a series of distinctive reheated or reactivated internal contacts appearing almost as bathtub rings, as mentioned in Chapter 3 (see Figure 3.2). From this extensive heating it is expected that the granite over vast areas in this region has been "cooked" enough to, in essence, reset or scramble its basic isotopic identity. Heating of this intensity and vastness can only be caused by prolonged heating by voluminous fresh high temperature magma such as that related to a major feeder or ascent zone (e.g., Marsh, 1989b; Wright & Marsh, 2016).

This extensively melted region in the Orestes Granite (an A-Type granite; Allibone et al., 1993a, 1993b) on the east side of Central Bull Pass (Figures 6.6 and 6.7), now called the Orestes Melt Zone, has been analyzed in some detail by Currier and Flood (2019). The melt zone here is ~20 m thick and several kilometers long and is characterized by two distinct melting facies, a distal zone (~9 m) and a proximal zone (~11 m). The distal facies is characterized by melting at lower temperatures and water-saturated, or near water-saturated conditions with near-eutectic melt compositions; miarolitic cavities are present (Figure 6.7). The facies proximal to the dolerite contact formed at higher temperatures and under

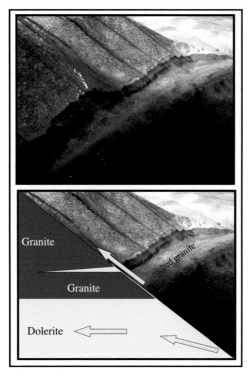

Figure 6.6 The splitting leading to a sill Doublet formation at the upper contact of the basement Sill in the east central wall of Bull Pass. (upper) The field relations showing the septa of melted granite between the two lobes; and (lower) an interpretation of the field relations indicating the possible movement of the magma upslope toward the Peneplain Sill.

water-undersaturated conditions. Proximal facies melting occurred in part by replacement reactions in restitic feldspars, resulting in producing plagioclase mantles on both restitic plagioclase and K-feldspar; these melt compositions diverge markedly from predicted eutectic compositions, tending toward enrichment in orthoclase components. Detailed thermal modeling indicates that this melt zone was active for a minimum of ~150 years, with a contact temperature of ~900°C, which is a time scale expected for filling of the Basement Sill. That this extensive melting occurred at this high level in the crust reflects the high temperature of these doleritic magmas and the duration of the magmatism (e.g., Bergantz, 1989).

At this original depth (~4–5 km) the initial crustal temperature before intrusion was perhaps ~150°C, and for this magma at ~1,250°C (see Chapter 10), the initial contact temperature would have reached about 750°C, but with prolonged passage of magma this reached ~900°C after ~150 years. Were this to occur deeper in the crust the amount of melting would be proportionally more voluminous, and it

Figure 6.7 Melting of granite by dolerite at the Doublet location in east central Bull Pass. (upper) The fine-grained complete transition from nearly fully molten granite to dolerite; (lower) the (mostly) fresh un-melted granite and another view of the contact where small miarolitic cavities are apparent in the melted granite.

almost certainly was. This melt zone, then, may well represent the top-most part of a much larger melt zone, telescoping outward and downward through the entire crust, as shown by Marsh (1984). It expands outward and downward because of the higher initial crustal temperatures, forming in essence a trumpet-shaped zone of partial melting, from which diapiric granitic melts may rise following the same path as the basalts that caused the melting (Figure 6.8). Some indication of a similar process operating in the Colorado Plateau has been deciphered by Putirka and Condit (2003).

Another dramatic indication of this intense heating by this local magmatism is found farther south along this same contact, up past the Granite Wedge. Displayed here is perhaps one of the most unusual and dramatic examples of a physical and chemical interaction between a mafic magma and partially melted granite ever discovered (Hersum et al., 2007). First some perspective: Although not specifically singled out previously, throughout this region are innumerable small granitic to aplitic dikes cutting primarily the immediately adjacent Basement Sill. In most areas these are actually small dikelets about 3–8 mm in width, often penetrating

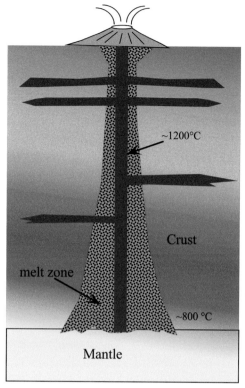

Figure 6.8 A schematic depiction of the establishment of an extensive melt zone adjoining a magmatic delivery column extending throughout the crust due to prolonged ascent of high temperature basaltic magma. The extensive melt at high level in the crust, as in central Bull Pass, increased progressively increased melting at depth where the crust is hotter (after Marsh, 1984).

50–100 m into the adjoining dolerite. Considering that dikelets of this size quench in less than ~1 minute, the propagation from the contact inward is near instantaneous; evidently a late-stage thermal contraction crack opens, and melt is vacuumed into it.[1] That melt is readily available is also remarkable. The largest of these granitic dikes is found in the area of Bull Pass where they commonly reach thicknesses of 20–25 cm and extend for 100s of meters. (A particularly large one, ~35 cm, cuts the base of the Dais Intrusion, as discussed in Chapter 7.) In central Bull Pass they form an indistinct extensive, somewhat orthogonal network reflecting a rather homogeneous, maybe isotropic local stress field upon emplacement. Although these larger dikes are in many places easy to trace, the

---

[1] Tiny granitic dikelets (~1 cm thick and 20 cm long) are found in many locations, even as far away as Solitary Rocks, back-veining into the dolerite at the contacts with granite, and, when first seen by Ferrar (1907), he took this as evidence of some granites being younger than the dolerites.

critical telling junctures at the sill contacts have proven, in spite of much effort, elusive. In this one place, however, on the high upland slopes above the valley, one such contact was unearthed, and it is most remarkable, indeed.

This is the upper lobe of the Basement Sill (the duplex or Feeder Zone area) and here there is a clear ~20 cm fine-grained, sound chilled margin all along the dolerite contact and the inner dolerite is medium grained. (On the opposite contact is a clear display of plagioclase comb layering.) Adjacent to this contact, on the granite side, is a similar bordering layer of granitic melt that was squeezed by compaction to form the heavily melted wall rock. While the dolerite was mostly solid but apparently ductile, this granitic melt broke open the chilled margin, very much as in opening a trap door, and penetrated tens of meters out into the dolerite (Figure 6.9). That the chilled margin could be deformed to this extent while still maintaining continuity signals a dramatic example of brittle–ductile deformation. The prolonged heating by the dolerite as it moved upward and outward southerly to the Kukri Hills melted the adjacent granitic wall rock outward by some tens of meters. The melting itself took place at high temperatures as it was at water-undersaturated conditions characterized by disequilibrium dissolution of dominantly quartz and alkali feldspar at temperatures of 900–950°C. The granitic melt was extracted from its residual matrix by lateral compaction and porous melt flow, collecting into a reservoir adjacent to the dolerite chilled margin. The coupled processes of compaction, melt extraction, and dike propagation were driven by the overpressure of dolerite emplacement, melting, and the contraction associated with emplacement cessation and cooling.

Should the dolerite sill exist in a mushy state deeper in the crust and become reactivated, the intrusion of this granitic dike back into the dolerite also offers ample opportunity for the magma to become isotopically contaminated (Figure 6.9). That is, if the dolerite undergoes further flow, the shearing motion will stretch out the granitic dikes, progressively thinning them and reducing the thickness to the extent that chemical diffusion, normally a very slow process, is greatly facilitated, thereby contaminating the magma in a characteristic pattern resembling the shear flow itself. A parabolic pattern of variation in initial Sr, perhaps resembling this process, has been found at a Portal Peak sill by Hergt et al. (1989a). A somewhat similar process of granitic melt production and possible basaltic magma contamination has been suggested by Heinonen et al. (2021), although the specifics of the actual contamination or homogenization processes are unclear.

A further clear sign of this deep feeder zone is the nearby presence just northwest of the two small frozen lakes in central Bull Pass, downslope from the duplex zone, of a piece of a coarse ultramafic intrusion made largely from Opx Tongue materials. Although the valley floor is mostly covered by erosional debris,

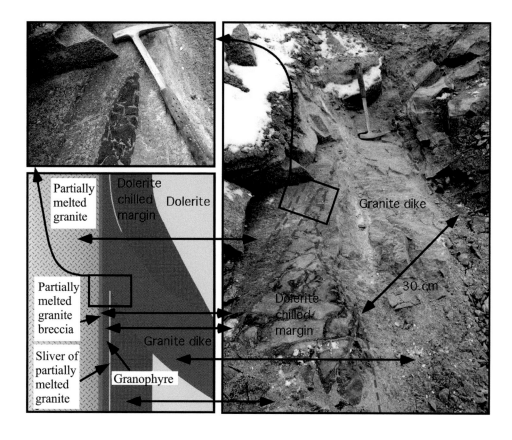

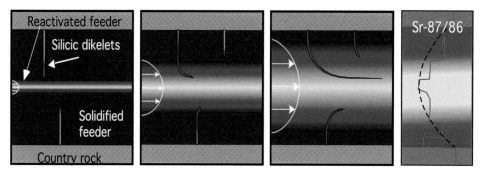

Figure 6.9 An extraordinary example of the formation of a granitic or aplitic dike in the south contact of the Feeder Zone, above Bull Pass, leading to formation of the southward-propagating Upper Lobe of the Basement Sill. The extensively melted granite wall rock (LHS), under compaction, has extruded a dike into the adjoining dolerite by ripping open – trap door fashion – the dolerite chilled margin (lower left depiction and RHS), extending some 100 m out into the dolerite (after Hersum et al., 2007). The lower series depicts a process whereby the granitic dike is assimilated into the dolerite, thus modifying its isotopic identity, through progressive reactivation of the magma undergoing shearing during flow.

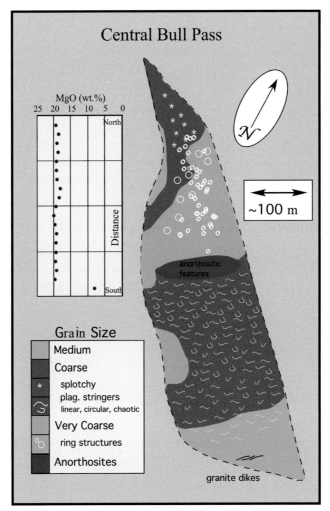

Figure 6.10 A field map of a part of the possible magmatic central ascent zone in central Bull Pass. The entire feature is packed by coarse Opx Tongue material (at ~20 wt.% MgO) with abundant vertical plagioclase schlieren structures and anorthosite layering. The surrounding area is covered by alluvium.

there is here an oblong window about 200 by 800 m exposing a coarse and highly varied ultramafic intrusion, which was mapped by pace and compass in terms of textural coarseness, and the presence and style of anorthositic stringers (circular, linear, and chaotic), and anorthosite features (splotches, layers, pods, and other segregations (see Figure 6.10)). The style of anorthositic features is related to the coarseness of the dominant Opx mineralogy, with the vertical circular or ring structures occurring in the coarsest rock and the medium grained rock having the fewest stringers. All of these characteristic features indicate vertical motion of the magma (Figure 6.5). Intermediate grain size rock, with highly grain-to-grain supported Opx, shows as spotted plagioclase or anorthositic concentrations, here called "splotchy." The larger geometry of the body is not possible to discern from the limited exposure, but the general impression is of a piece of a much larger body where the medium grained rock, which is as coarse as usual Opx Tongue rock, is near the margin and the coarsest rock is nearer the throat of the body. The analyses of a series of 20 samples taken from north to south shows them all, excepting one, to have an MgO content of nearly 20 wt.% (Figure 6.10). The exception at 7.5 wt.% is at the outer margin and is medium-grained rock.

### 6.3 Bull Pass to Wright Valley to the Dais

In sum, all evidence points to the magmatic throat of this system as being in central Bull Pass with major and minor lobes of the Basement Sill moving outward in all directions (Figure 6.11), perhaps similar to inflation filling of large pahoehoe flood basalts (Self et al., 1996, 2000; Anderson et al., 1999). Moreover, the large dike-riddled granite block at Orestes Valley has apparently settled into the underlying Basement Sill, depressing the contact (Figure 6.12). Although the contact in Orestes Valley is covered, this feature, along with the others already described in this area, suggests that this block may be the deeper root of an overlying caldera system. The Orestes block has slid, rotated, and sunk into the sill magma, indicating a structure that may have continued upward higher into the crust as the

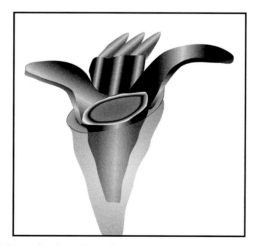

Figure 6.11 A schematic depiction of the style of emplacement of the Basement Sill emanating from the ascent zone in central Bull Pass as a series of outward extending and coalescing magmatic lobes

Figure 6.12 East wall of Bull Pass, near Mt. Orestes. The large granite stoped block sinking into the Basement Sill, perhaps representing the lowermost part of an upward extending caldera-like system. (upper) The form of the block and the form of Orestes Valley suggests magma may have ascended upward along each side of the block. (lower) Notice the large fragment of granite undergoing incipient stoping.

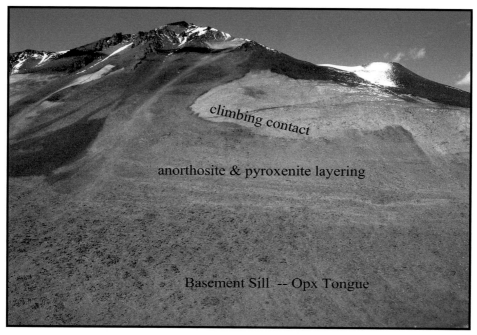

climbing contact

anorthosite & pyroxenite layering

Basement Sill -- Opx Tongue

Figure 6.13 The Basement Sill beneath Mt. Odin in the south wall of Wright Valley. Here the sill climbs upward to quench (inferred) against the overlying Peneplain Sill, forming a relatively thin layer on the dip slope, while at the same time propagating deeper westward toward the Dais and upper Wright Valley. The lumpy brown lowermost section is typical Opx Tongue material and the upward transition, showing distinctive banding and layering, reflects incipient ponding, crystal sorting and settling, and delicate layer formation; the dark pyroxene layers form distinct brows on the slope. (Running diagonally upward across this face, westerly from the valley floor, there is a faint, but still distinct and well-trodden foot path apparently connecting what was once a NZ Geological Survey camp at the east end of Lake Vanda to a remote high camp in the Asgard Range, perhaps near Mt. Freya. This era of camps in Wright Valley is explained by Harrowfield (1995).)

whole sill system was being established. Other evidence of this may be exposed in the south wall of Wright Valley.

Across from Bull Pass in the south wall of Wright Valley, the Basement Sill appears as a series of southward propagating coalescing lobes of dolerite (Figures 5.8 and 6.13). The upper contact climbs strikingly upward, apparently intending to form an eruptive vent, then quenches against the overlying Peneplain Sill. This forms a classic climbing contact showing a distinctive curtain-like coating of quenched margin on the valley wall slope. This feature coincides on strike with the climbing contact at southwest Bull Pass and with the major ascent zone in central Bull Pass. This may collectively represent a major fissure ring fault associated with this same deep-seated caldera system. Moreover, that this may

Figure 6.14 Lower section in the layered zone of the Basement Sill beneath Mt. Odin on the south wall of Wright Valley. Two examples of small grains of plagioclase draining or sieving from coarse orthopyroxenite forming flame structures leading to felsic-anorthositic pyroxenite.

have been floored by a major mass of pooled magma is exemplified by an abundance of distinct large-scale layering along this whole section westward between the areas of the Comrow Glacier, Mt. Odin, and beyond. The section here is floored by typical massive, lumpy Opx Tongue material with the layering appearing in the transition to the higher upward climbing contact material.

This layering is in contrast to small-scale, more haphazard layering common to many areas of the Opx Tongue. The layering appears as light and dark horizontal bands and is most apparent from the air. On the ground, although less distinct, the layering is still obvious as horizons of massive pyroxenite brows or ledges separated by felsic or plagioclase-rich horizons. Here and there within the felsic horizons are delicately layered meter-scale patches rich in anorthosite; small grains of plagioclase appear to be draining (sometimes mimicking flame structures) from the overlying coarse pyroxenite (Figure 6.14). This horizontal layering becomes much more prominent to the west as the sill dips strongly northward toward the Dais and becomes covered by scree.

This part of the sill on the south side of Wright Valley has been numerically modeled by Petford and Mirhadizadeh (2017) in two and three dimensions employing the exact form of the sill, as displayed in the valley wall and using a

Figure 6.15 The Dais Intrusion in central Wright Valley looking southwest. The massive lower section is coarse-grained Opx Tongue rock, the middle section shows distinct anorthositic layering on many scales, and the upper section is mainly felsic orthopyroxenite. The upper contact on the right (west) is a climbing contact, with the dolerite eating into and disaggregating the overlying granite. The lower contact is buried and, judging from the pattern of stratigraphic MgO concentration, it may be thicker by ~300 m or more.

particulate rheology appropriate for a crystal-rich slurry magma. Among a host of useful results, they show that, during the lateral shear accompanying emplacement, irregularities in the lower contact promote local eddies, perhaps leading to the formation of ultramafic pods and, in response to the dilatancy imposed by the intergranular pressure, significant layering may form. Although they model the flow being driven from east to west, the actual emplacement direction may well have also had a strong north–south component, which would encourage similar effects.

The pervasive layering here intensifies to the west, culminating at the east end of the Dais as several pronounced thick (~10 m) white bands obvious from a distance of 10 km (Figure 6.15). On the ground there is clean, clear anorthosite layering of every scale from meters to millimeters, principally marking specific horizons, all of which involve various degrees of sorting between coarse (1–25 mm) pyroxene, principally Opx but also subordinate Cpx, and also fine grained (~0.1 mm) Opx and plagioclase. This is the singularly remarkable Dais Intrusion.

# 7

# Dais Layered Intrusion

## 7.1 Introduction

This relatively small intrusion has many of the features of large ultramafic, layered intrusions. And it has two critical principal features not found elsewhere that make it of singular importance: Its mode of formation, by ponding as a crystal-laden magma, is abundantly clear, and, due to its size, it cooled quickly enough to capture critical ongoing crystallization and layering processes rarely, if ever, previously seen. The overall body is described first, along with the basic types and the styles of layering; this is followed by the general whole-rock stratigraphic compositions and petrogenetic discrimination diagrams relating to crystal fractionation and differentiation. Above all, it is perhaps the unusual captured or quenched petrographic textures that also make this intrusion of singular importance, which closes this chapter.

## 7.2 Scope of the Dais Intrusion

The exposed body, which is about 550 m thick, consists overall of several thick (~5–10 m) anorthositic horizons within massive pyroxenite. The layering is approximately horizontal with a slight (5–10°) southwestward dip and a variable strike from E–W to SE–NW. (This attitude may reflect the regional tilt.) The upper contact against granite is well exposed and the lower contact is covered and, judging from the form of the MgO profile in comparison with other profiles through the Basement Sill (e.g., West Bull Pass; see Figure 5.6), the total thickness could be about 900 m.[1] The upper contact is a climbing or reactivated contact, as defined in Chapter 3, and is transitional into disaggregated or "digested" granitic wall rock. The base of the body is massive coarse high MgO (~20%) pyroxenite

---

[1] A partially successful effort was also made to seismically source the nature of the buried basal section by Drew Feustal (2005, personal communication).

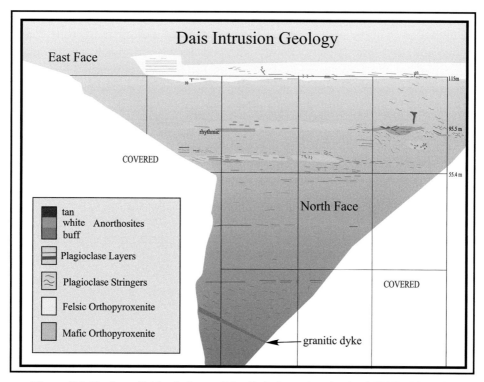

Figure 7.1 Geology field relations of the Dais Intrusion (revised 2020). There is nearly every scale and type of anorthositic layering, which, although discontinuous, extends tens of meters at all specific horizons with knife-sharp edges and constant thicknesses. The lightness of the blue coloring depicts the abundance of plagioclase and the tan to orange on the upper right is the area of dry-grained sorting (see text for more description).

with abundant horizontal plagioclase stringers and the top is low MgO (~7%) dolerite.

In the detailed sketch map of the geology (Figure 7.1) there are three basic units: massive orthopyroxenite, concentrated orthopyroxenite, and anorthosite[2]. The anorthosite is further broken down into three varieties: tan, buff, and white, which are indicative of the content of fine-grained Opx. That is, the anorthosite layers are actually made up of concentrations of fine-grained (~1 mm) plagioclase and orthopyroxene. The tan anorthosite, which in the field appears very much as sandstone, is a 50:50 mix of plagioclase and Opx. The white

---

[2] These might be more universally called, respectively, websterites, gabbro-norites, and anorthosites, but the field terms used here are much more descriptive and convenient and are in keeping with all prior designations for these rocks.

anorthosite is almost pure (>~90 vol.%) plagioclase, and the buff anorthosite is a combination of these endmembers. The concentrated orthopyroxenite is coarse pyroxenite that has been evidently stripped locally of much of the usual interstitial small plagioclase. In strong contrast to the pervasive fine plagioclase, the fine Opx is not obvious in the normal concentrated Opx Tongue rock, and the unique origin of this material at this magmatic location is addressed in Section 7.5.

The lower half of the Dais is typical massive Opx Tongue cumulate laced in its lower half by pervasive plagioclase stringers. Within this sequence are two thin (~20–40 cm) horizons of prominent anorthositic layering. This is a horizon because the layering is discontinuous along strike and comes and goes as well-formed layers at distinct levels perfectly on strike for 50 m or so until covered with debris (see Figures 6.14, 7.1, and 7.2). The coming and going of these strong layers give the impression of lobes gently cascading down dip, gathering plagioclase from the overlying coarse Opx as they travel. The lower of these two layers (~30–40 cm) caps the lower zone of pervasive plagioclase stringer development and in the western part becomes a doublet layer with a second weaker layer immediately (~1 m) below the principal layer. The upper half of this massive cumulate section has few plagioclase stringers but contains many thin (5–15 cm) discontinuous distinct anorthosite layers scattered throughout this section. The second of the two thicker and well-formed layers mentioned occurs about midway in the upper half of this section.

The upper half of the Dais contains the thick light-colored bands noticeable from afar. Each of these three layers is ~10–15 m thick and, again, are not single continuous layers but are, in effect, event horizons containing a vast assortment of styles and scales of layering at specific horizons (Figures 6.14, 7.1, and 7.2). Of these three event horizons, the upper layer is more typically a thick felsic zone containing thin (5–10 cm) discontinuous anorthositic layers ~1–3 m in length. It is within the other two event horizons that most of the great variety of layering is found, including the tan, buff, and white anorthositic layers. Associated with these anorthositic layers are patches of unusually massive and dark pyroxenite (Figure 7.1); some of these are intimately associated within anorthosites and others, the largest of which occur immediately above the tan, buff, and white anorthosites. The largest of these (see the western end of the middle event horizon of Figure 7.1) is somewhat trough or funnel-like and traverses 4–5 m downward just above the principal horizon itself. The other characteristic form of layering is well-formed repetitive or rhythmic layering with a wavelength of 3–5 cm that persists upward for 2–3 m and laterally for 10–15 m (Figure 7.2). This is highly reminiscent of the well-known inch-scale layering found in many layered intrusions.

Figure 7.2 Three styles of anorthosite layering in the Dais Intrusion: (upper) fine scale rhythmic layering in the upper section; (middle) coarse discontinuous layering in the middle section; and (lower) coarse plagioclase stringers in the lower section

## 7.3 Basic Compositional Variations

In order to define the chemical nature of the body and reveal possible cryptic layering, the Dais Intrusion was sampled upward at 5 m intervals without regard to layering; characteristic samples of the layering were taken separately on multiple scales. The variations in whole rock CaO and MgO, superimposed on a field scenic

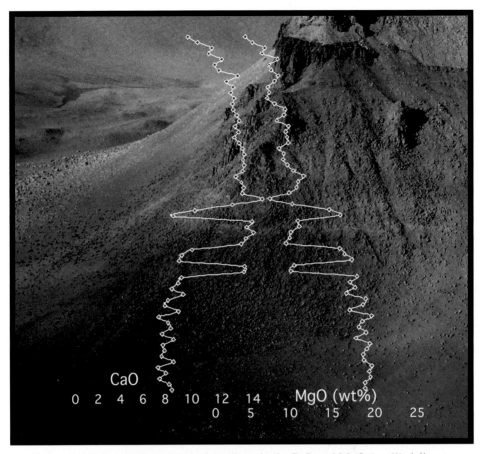

Figure 7.3 Dais Intrusion chemical stratigraphy in CaO and MgO (wt.%) delineating the principal layering event horizons and the antipathetic relation between MgO and CaO, which are representative of the populations of Opx and plagioclase.

(Figure 7.3), clearly show the correlations with the layering, and Figure 7.4 is marked with lines emphasizing the striking chemical differences between the lower and upper sections. The lower section is ultramafic orthopyroxenite with MgO at ~20 wt.%, whereas the upper section is distinctly tholeiitic at MgO ~7%. In the chemical stratigraphy (Figures 7.3 and 7.4), the three anorthositic horizons are clear, with the middle horizon being the thickest (~25 m); across each horizon MgO varies by about 10 wt.% and CaO by about 5 wt.%. At the top of each horizon is an abrupt transition to a massive pyroxenite erosional brow. The lower massive section exhibits variations over successive 5 m samples of 1–4 wt.% in MgO and 0.5–2 wt.%. in CaO. The top or felsic section shows similar variations, although as a whole the compositional variations are much smoother.

Figure 7.4 Dais Intrusion chemical stratigraphy with representative straight lines (red) and curves (blue) emphasizing the compositional contrasts between the upper tholeiitic section and the lower ultramafic pyroxenite part of the body, with an intermediate transition oscillating between these two end members. The middle section, or event horizon, represents a major transition in crystal sorting and fractionation, leading to a major differentiation event.

The intrusion as a whole can be viewed as three gross units each of a thickness of 100–150 m: A lower ultramafic orthopyroxenite unit with MgO ~20 wt.% and CaO ~7 wt.%; an upper tholeiitic unit with MgO ~8 wt.% and CaO ~13 wt.%; and a thick transitional unit containing the major layering event horizons, oscillating between the lower and upper units. At the very top is a thin zone of disaggregated digested granite. The sequence can also be depicted as smooth variations in composition through the top and bottom units, with a sharp, step-like transition in the middle. Starting at the bottom, the composition varies smoothly (aside from the so-called cryptic variations) upward to the last major layering event horizon and then steps to the smooth curve through the upper tholeiitic section (see Figure 7.4).

On the other hand, starting from the top the composition varies smoothly until reaching the lowermost layering event horizon where it makes a sharp step to the lower ultramafic sequence. Although these compositional depictions are not intended to suggest a specific physical or chemical process, they do show that the overall compositional variations within the intrusion undergo a fundamental change in character in the transition through the section containing the layering event horizons.

This is all further emphasized by considering the same data in a plot of whole rock CaO versus MgO (all in wt.% in Figure 7.5a). The upper tholeiitic and lower ultramafic sections show a strong chemical separation with the rocks of the layering events divided between the two groups. That is, the anorthositic layers fall in with the upper section rocks and the pyroxenites fall in with the lower section rocks. The effects of the "digestion" of the granite roof rock at the upper contact is also clear in defining a third group of compositions showing a separate distinct trend normal to that of the main sequence, depicting a mixing line between the upper sections and the granite. Much of the general trend of compositions, especially in the lower section, can be portrayed as a simple mixing sequence (mechanical and chemical) between the heavily pyroxene-rich sample A-283 in the lower section and the plagioclase-rich sample A-357 from the uppermost of the three layering event horizons. Knowing the exact *spatial* location of each sample throughout this section allows the chemical variations to be ascribed to specific changes in the mineralogical or physical character of the rocks themselves. Over a 15-m stratigraphic distance, four samples (A-354, A-355, A-356, A-357) display a continuous change in composition (from 15 to 7 wt.% MgO), which defines a large part of this entire range in rock composition. And two successive samples just 5 m apart (A-355 and A-356) define the full compositional gap (14–9 wt.% MgO) between the lower ultramafic and the upper tholeiitic rocks. It is essential to appreciate that local physical processes involved in sorting and separating the large pyroxene grains from the small plagioclase grains completely control this spatial variation in composition. That is, both the cause and the effect can be directly ascribed to specific physical processes and characteristics seen *spatially* directly in the rocks themselves. The fact that these are exact spatial variations as well as chemical variations is important to appreciate.

This simple compositional dependence on pyroxene and plagioclase is particularly pronounced in a Pearce-style molecular plot depicting the tradeoff in pyroxene (0.5(Mg+Fe)) and plagioclase (2Ca+Na) control (Figure 7.5b). The same four samples spanning 15 m stratigraphically nearly define the whole trend between pyroxene and plagioclase. In striking contrast is the strong "feldspar" influence shown by the samples from the upper contact area involved in digestion

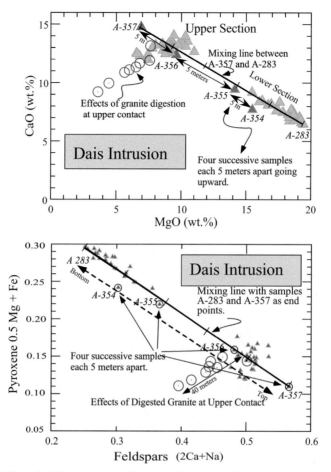

Figure 7.5 Chemical fractionation diagrams for the whole rock compositions of the Dais Intrusion. (upper) CaO vs MgO illustrating the major fractionation/differentiation effect of four successive samples, each 5 m apart, spanning from the ultramafic pyroxenite at the base to the tholeiitic section at the top, and also showing an effective mixing line between two samples (A-357 and A-283) illustrating the effects of simple crystal sorting and fractionation between these two end members. Notice the separate distinctive trend of the partially digested granitic wall rock from 40 m at the upper contact. (lower) The same bulk rock data plotted as molecular Pyroxene (0.5(MgO + CaO)) and Feldspar (2Ca + Na) illustrating the tight controls that each of these components have on the overall spatial pattern of differentiation. The four spatially successive samples, each 5 m apart, and the effective mixing line due to crystal sorting is also shown, as are the samples contaminated by ingesting granite at the roof.

of the granite roof rock, which produces a trend normal to that of the main crystal sorting sequence. It is fundamentally important to observe these specific processes of crystal sorting in the rocks themselves, which can be seen here in perhaps unprecedented spatial clarity.

## 7.4 Petrographic Textures

Compared to large, layered intrusions, the Dais Intrusion is small, and its solidification time was therefore relatively rapid (~2,500 years). Textures normally erased through annealing during long cool-down times are preserved here. This is particularly apparent in the preservation of the coarse pyroxene and especially the fine, even sugary, granular texture of the interstitial plagioclase at all levels within the Dais Intrusion stratigraphic sequence.

### 7.4.1 Orthopyroxenites

The typical rocks themselves (Figure 7.6) show the basic granular styles of these Dais orthopyroxenites. It cannot be emphasized too strongly that these basic working materials produce all the compositional variations and, more importantly, the intimate textures captured by fairly rapid cooling. The dominating feature of these rocks, as repeatedly emphasized, is the pervasive presence of the large (5–20 mm) grains of pyroxene, mostly orthopyroxene, which vary in abundance

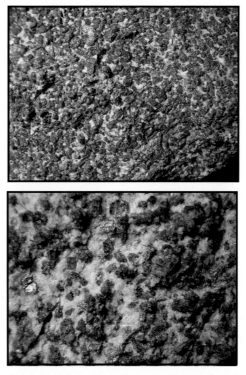

Figure 7.6 Typical pyroxenitic rocks of the Dais Intrusion. (upper) Ultramafic orthopyroxenite with high concentration of close packed Opx Tongue crystals (field of view ~10 cm); (lower) felsic orthopyroxenite, a mixture of coarse Opx and sugary plagioclase (field of view ~4 cm).

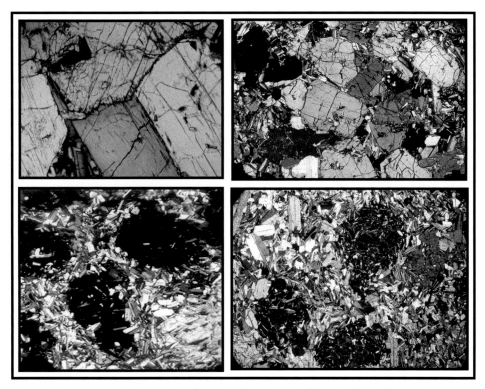

Figure 7.7 Photomicrographs of Dais rocks. (upper left) densely packed orthopyr-oxenite clot showing approximate characteristic 120° grain boundary fabric (field of view 1 cm); (upper right) slurry of individual pyroxene crystals, mainly disaggregated, but some still attached, amongst a sea of small plagioclase (field of view ~1.5 cm). (lower two views, each ~0.75 cm) Each exhibits small grains of plagioclase sieving downward through a sieve matrix of orthopyroxene crystals. The Opx grains show similar optical orientation (extinction) because they are long in their c-axis and are commonly oriented lengthwise and parallel in the anorthositic layer.

from highly concentrated (~55 vol.%) pyroxenite (Figure 7.6a) to felsic pyroxenite (Figure 7.6b), where 10–25 vol.% pyroxene is scattered within a matrix of small, sugary (~0.1–1 mm) plagioclase. In thin sections (Figure 7.7), the concentrated pyroxenite (Figure 7.6a) is a collection of large grains crowded together with small plagioclase sequestered along grain boundaries (there are some older plagioclase of similar size within the Opx grains) or as clusters of large grains with plagioclase-free boundaries that form clots with inter-crystal triple junction boundaries approaching 120° (Figure 7.7). The pyroxene grains themselves range from idiomorphic to ragged and cracked grains with characteristic undulatory extinction. Throughout these rocks the large variations in crystal size (plagioclase to pyroxene) show little sign of massive annealing.

The felsic pyroxenite (Figure 7.7) has two distinctive features: First the large rod-like grains of pyroxene are commonly oriented with their c-axis in the plane of the layering; second, the small interstitial plagioclase grains appear to be draining downward from the matrix of pyroxenes that acts as a well-structured sieve. The long axis of each plagioclase lath becomes increasingly vertical with an approach to the slot between adjacent pyroxene grains. This is especially common in samples immediately above any anorthositic concentration, be it a small stringer or one of the types of well-formed sandstone-like anorthositic layers. And ensembles of nearby pyroxene crystals commonly exhibit almost exactly the same extinction angle over areas as large as several thin sections covering tens of centimeters to meters. It is as if the draining or sieving process also highly influences the orientation of the structural members (i.e., the pyroxene) of the sieve itself. This texture suggests that much of the layering has formed locally by this process, which is known as kinetic sieving and will be discussed in more detail henceforth.

### 7.4.2 Anorthosites

Regardless of where they are found, the anorthosites are almost universally fine-grained, sugary rocks. They form the most delicate structures. Sinuous stringers, crenulated layers, layers with rippled surfaces, and massive (1–2 m thick and 10 m long, Figures 7.2 and 7.8) "sandstone"-like areas are some of the many styles of layering exhibited by the anorthosites. There are four fundamental characteristics of these rocks that are notable.

First, they are unusual in consisting not only of plagioclase but also of significant concentrations of small sugary orthopyroxene. This is unusual because, although small grains of plagioclase are common in all rocks, small grains of orthopyroxene are not at all obvious in any of the basic sill rock types. Small fresh crystals of orthopyroxene are especially absent in the pervasive coarse-grained Opx Tongue rocks. These small Opx grains are similar in size, but more equant in shape, than the coexisting plagioclase. The modal concentration of orthopyroxene determines the color of the anorthosite which is found in three varieties: white (<10 vol.% Opx), buff (Opx ~25%), and tan (Opx ~50%). The origins of these three varieties is not entirely clear.

Second, within the tan anorthosites (i.e., 50:50 opx:plag), collections or aggregates of grains of orthopyroxene have "annealed" into large (~100 grains) optically continuous *patches* enclosing the small grains of plagioclase (Figure 7.8). These patches resemble inclusion-riddled alumino-silicates in pelitic hornfelses. Around the edges of these contiguous patches small satellite grains of orthopyroxene can be seen here and there, also having this same optical orientation indicating that they are evidently already connected to the central mass

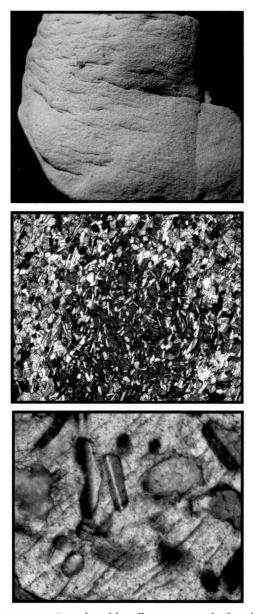

Figure 7.8 Orthopyroxene "porphyroblasts" spontaneously forming in tan "sand-stone" anorthosite. (upper, 20 cm) Typical rock consisting approximately of a 50:50 vol.% mix of sugary orthopyroxene and plagioclase; (middle, 10 mm) patches of orthopyroxene forming through annexation and annealing of small grains, notice the small nearby archipelago of grains of common optical orientation; (lower, 1.5 mm) tiny plagioclase grains enclosed in orthopyroxene becoming rounded upon undergoing apparent resorption.

in the third dimension (Figure 7.8). Within this halo, however, most Opx grains do not yet share the optical orientation of the patch. These patches clearly represent the nascent development of large single grains of orthopyroxene, which adds a sieve-like structure to the otherwise homogeneously granular assemblage. And that they are routinely found in random cuts suggests they are also extensive in the third dimension (i.e., rod-like) similar to the more mature Opx phenocrysts found elsewhere. That these nascent phenocrysts or igneous porphyroblasts have begun sieving the plagioclase can also be seen by the arcuate pattern of plagioclase grains included near the margins of the growing – by annexation – porphyroblasts (Figure 7.8b). Moreover, in the centers of the Opx patches the initial identity of the original individual small orthopyroxene grains has been completely lost, having been collectively annealed into a single large, optically continuous Opx grain (Figure 7.8c). And the included small grains of plagioclase are themselves, remarkably enough, beginning to disappear; the corners of the included plagioclase grains have become rounded, and some seem to have mostly disappeared. It is, clearly, as if the plagioclase has begun to dissolve due to diffusion (Figure 7.8c). The well-known sluggishness of diffusion of plagioclase components within a solid pyroxene matrix is well appreciated here, but nevertheless something of this nature is obviously taking place. (How does it similarly happen in alumino-silicates?)

Third, in outcrops of mainly the white anorthosites (Opx <10%), large (2 × 15 cm) patches of *massive* grey plagioclase are found sporadically at some horizons (Figure 7.9). These patches of massive grey plagioclase are particularly distinctive and peculiar because the anorthositic rocks are generally always fine-grained, sugary granular rocks. In thin sections, these massive patches consist of aggregates of individual grains of slightly coarser plagioclase quenched in the process of annealing into single large, optically continuous crystals (Figure 7.9b). Individual grains of different optical orientation, and thus different twinning, have been captured in all stages of annealing into single large grains. And ghosts of the original individual grains are in some places discernable within single large grains. That a local minor fluid phase may have sometimes facilitated this process may be inferred from sporadic occurrences of tiny patches of sericite.

The fourth distinctive petrographic feature of the Dais rocks involves the occurrence of older large grains of primary orthopyroxene caught up within anorthosite layers. Large grains of original orthopyroxene, similar to those populating the Opx Tongue and presumably residual from the sorting process, become increasingly ragged and diffuse with distance inward into anorthosite from an adjoining layer of pyroxenite (Figure 7.10). Near the interlayer boundary, individual grains are ragged but show an idiomorphic outline with good optical continuity. With distance (5–10 cm) deeper into the anorthosite, Opx phenocrysts

Figure 7.9 Hand sample example of massive plagioclase rock (20 cm), and three views of the process of multiple individual grains annealing into large optically continuous patches of plagioclase (field of view in each is 3 mm)

become increasingly ragged to the point that only bits and pieces of the original crystals remain; yet the residual pieces remain in optical continuity and enough of the rims remain to discern that the original crystal size was similar to those of the adjacent layer of pyroxene. Deeper in the anorthosite layer there are very few if any tiny residual fragments of these pyroxenes. It appears as if original pyroxene phenocrysts caught in anorthosite layers are unstable and undergo progressive dissolution with time and distance into the anorthosite layer. This may be an example of the process proposed by Boudreau (1995, 2011; Boudreau & McBirney, 1997) wherein crystals mechanically placed in a medium of contrasting chemical equilibrium dissolve away. That is, the prevailing ensemble of governing chemical potentials is controlled by the largest mass of any single phase, which here is clearly plagioclase, making Opx unstable. A similar process occurs when clastic sediment undergoes high-grade thermal metamorphism; some phases thrive at the expense of others.

To further understand the possible influence of annealing or ripening in the maturation of the Dais textures leading to the remarkable layering, a single layer in

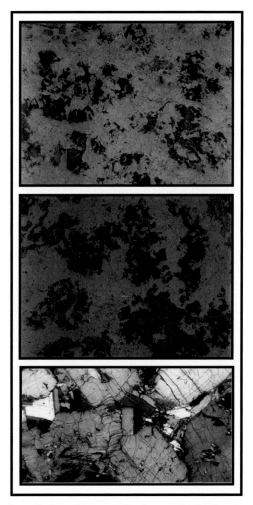

Figure 7.10 A series of photomicrographs from a bed of orthopyroxene (lower) going 20 cm upward into an anorthosite layer where the Opx becomes increasingly ragged and fragmental, although still optically continuous, as it undergoes progressive dissolution (field of view ~1 cm)

a larger sequence of rhythmic layering was imaged using X-ray computed tomography (CT) by Jerram et al. (2010). This procedure allows the texture to be quantitatively analyzed at high resolution in two- and three-dimensions where the ongoing growth of the original primocrysts can be evaluated relative to active compaction, evacuation of interstitial melt, and secondary crystallization. The initial cumulate texture was formed by a framework of Opx (37–47 vol.%), which in time overgrew by an additional 7–10 percent. At the same time Cpx and plagioclase nucleated and grew in the interstices as fine micro-crystals as the interstitial melt was expelled upward. This upward migration was systematically arrested due to rapid growth of Opx and Cpx dendrites, forming oikiocrysts enclosing the initial microcrystal population of existing plagioclase nuclei, along with plagioclase overgrowths in the areas where the pyroxenes were absent. When coupled with the rate of cooling at this place in the intrusion, this recrystallization event took perhaps about 140 years. This represents, in essence, a microcosm of the ongoing processes taking place on many scales throughout the Dais as interstitial fluids were being expelled upward in response to various styles of deposition, sedimentation, and crystal sorting.

These processes are also reflected in the detailed mineral chemical stratigraphy throughout the intrusion (Bedard et al., 2007). That is, one of the most striking features of the mineral–chemical stratigraphy (see figures 15–17 in Bedard et al., 2007) is the extremely modest range of cryptic variation in mineral composition throughout the body over some ~400 m. The maximum and average Mg-number for Opx and its $Cr_2O_3$ content are uniformly high in all lithologies, which are correspondingly similar to those observed in the chilled margin micro-phenocrysts. Plagioclase also shows very little systematic cryptic variation in An-content (~$An_{85}$), essentially regardless of rock type, which is similar to that of chilled margin euhedral micro-phenocrysts. This is somewhat in contrast to the variation in Opx composition across the Basement Sill in West Bull Pass, which is likely a reflection of the effects of density sorting during flow differentiation, to be presented in more detail later. It seems that the process of ponding and successive slurry deliveries along with a wide assortment of crystal sorting processes produced a compositional package somewhat distinct from that of simple shear flow in the sill itself. The final assemblage more closely resembles a true magma chamber rather than simply a thick sill.

In this regard, the scale of pyroxenite–gabbroic cyclic layering in the Dais is remarkably similar to that found in the Bushveld Complex. The mechanism responsible for this has been suggested to be by retention of plagioclase in the melt, perhaps due to neutral buoyancy, while pyroxene is being deposited, which is then followed by plagioclase deposition, over and over (Eales et al., 1991; Cawthorn, 2002). Although perhaps possible, with the extensive late stage annealing in the Bushveld, this delicate process has been difficult to validate, but the abundant evidence in the Dais adds a fresh perspective as to how such layering and texture development can routinely take place by relatively simple processes within a thick dense slurry or mush. The observations by Eales et al. (1991) are especially pertinent as they describe the vestige phase of this process of plagioclase incorporation and dissolution included in both Opx and olivine and perceptively interpret the textural sequence to represent (p. 484): "(b) inclusions represent plagioclase nucleated within a bottom zone enriched in the components of feldspar, consequent upon in-situ growth of pyroxenites on the floor of the complex; (c) the texture is the result of subsolidus recrystallisation, or solution and reprecipitation in crystal mushes, as described by Hunter (1987)" (Eales et al., 1991, p. 484). That the much larger Opx crystals may themselves have also formed by a process of aggregation and rapid annealing is much harder to perceive, but the final result is certainly very much what might have happened in the Dais rocks if the cooling had taken much longer.

## 7.5  Summary of Dais Characteristics

Periodic shallow cascading slurries moving southwesterly from Central Bull Pass, containing mixtures of coarse old grains of pyroxene and some plagioclase and small fresh grains of plagioclase nucleated during ascension, have formed the Dais sequence. Well-formed layers on all scales have formed from the physical sorting of small (~0.1 mm) grains of plagioclase and enormously larger (5–10 mm) grains of mainly pyroxene. This sorting has occurred by a number of processes, the most prominent of which is simple shear, causing dilatancy, accompanied by sieving during large scale settling and compaction with upward expulsion of melt. The extreme disparity in grain size allows tiny plagioclase grains to settle or drain, almost as a fluid, through concentrated regions of pyroxene, which have acted as interconnected sieves, to collect at convenient locations into anorthosite layers (see Figures 7.11 and 7.12). This physical process has been termed kinetic sieving by Middleton (1970, 1993) and the physics of this and similar processes have been investigated in some detail by, among others, Savage and Lun (e.g., 1988), Jaeger and Nagel (1992), Jaeger et al. (1996), Major and Pierson (1992), Nedderman (1992), Knight ct al. (1993), Vallance (1994), and Pouliquen et al. (1997). Because of the obvious possible importance of this process to these rocks, more on these processes will be treated in more detail next.

The sorting process mechanically juxtaposes crystals that, prior to complete solidification, begin to chemically re-equilibrate upon final deposition. Where crystals are of a similar size and composition (e.g., sugary plagioclase and orthopyroxene), coarsening is the first obvious form of re-equilibration. Tiny crystals are unstable relative to large crystals, due to the higher Free Energy of high curvature grain boundaries, and coarsening is strongly promoted. Orthopyroxene apparently more easily forms aggregates enclosing plagioclase, expanding outward by annexing the next-nearest neighbors. The aggregate interior anneals quickly into a single large optically continuous crystal while the enclosed plagioclase grains seem to dissolve away, in much the same way, apparently, as the large Opx grains do when embedded in a plagioclase dominated medium. (Just where the dissolving plagioclase components go is not readily apparent, perhaps going into solution with the interstitial basaltic melts.) The new large pyroxenes mechanically

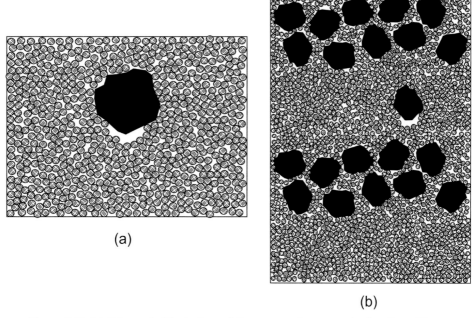

Figure 7.11 (a) Schematic illustration of the progressive upward migration of large grains in a shearing or shaking assemblage of smaller grains by small grains continually rolling beneath large grains and steadily lifting them. (b) A layer of collected large grains acts as a sieve, allowing small grains to settle through, but large grains are trapped in the layer or collectively migrate up to another layer.

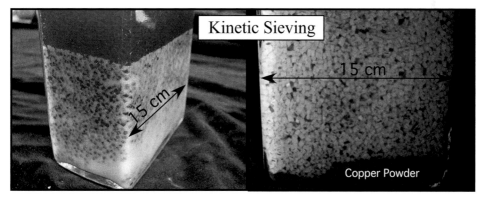

Figure 7.12 (left) A water-based slurry undergoing kinetic sieving of small sand grains (white) through a matrix of coarser grains (red); (right) when the grains become smaller and denser (copper powder), the process of kinetic sieving becomes more pronounced, and in a tall column a point may be reached where the matrix becomes clogged with sinking small grains and a layer will spontaneously form

act further to sieve the interstitial plagioclase grains into increasingly cleaner layers of anorthosite and pyroxenite. The massive areas of plagioclase also seem to grow in a similar fashion, but only where the medium is relatively pure in plagioclase. Given a significantly longer cooling time, these processes would soon run to completion and produce a final coarse-grained texture that would reveal little of the initial state, processes of sorting, and subsequent stages of textural re-equilibration.

The great abundance of fine-grained orthopyroxene crystals, with a size similar to the plagioclase, is surprising, as it is not found elsewhere in the Basement Sill. This is another reflection of the attempt to achieve chemical equilibrium between the basaltic melt and the huge mass of entrained Opx crystals. That is, the large mass of Opx being bathed in the basaltic melt literally forces the melt to become saturated in Opx. This is similar to the process of so-called salting an experimental charge in a particular phase to encourage or spawn nucleating and crystal growth. (A common novel practice to achieve saturation with olivine in basaltic melt experiments is to make the enclosing capsule out of olivine dominated material, like dunite.) This bathing produces massive nucleation of orthopyroxene, which is nearly everywhere observed in the Dais and associated rocks along the south wall of Wright Valley.

## 7.6 Sieving and Sorting Processes

A highly revealing demonstration of the dynamics of sorting between small and large particles is the widely appreciated experiment whereby a large (~2 cm) steel ball bearing is placed near the bottom of a column of common dry beach sand. A steady vibration of the column induced by tapping on the sides will eventually culminate in the appearance of the steel ball at the surface. The reason for this is that, during any slight deformation, small gaps will appear beneath the ball into which will roll sand grains. The gaps will never be large enough to allow the large ball to settle and over time the ball is steadily lifted until it reaches the surface (Jullien et al., 1992; Cooke et al., 1996). A similar process occurs in mudflows or lahars when large boulders are elevated to the surface. And on a more everyday level, when caught up in a slow-moving traffic jam, the easiest way to change lanes is in front of a large truck. Large trucks need a significant open space or cushion in front to assure enough space to stop, and this space is usually more than adequate to fit a passenger car into. The net results in some instances when enough cars do this is that the truck actually moves backwards relative to the bulk flow of traffic.

This process involving Opx, and plagioclase grains is illustrated by Figures 7.11 and 7.12, where this also shows an ensemble of large grains have collected at a particular horizon to themselves form a sieve. Once a group of large grains encounter one another, inter-grain hindrance retards further travel, and a layer is formed. Moreover, during shear flow these large grains are similarly lifted while the small grains settle to the bottom, regardless of the density contrast, as long as they are each denser than the enveloping liquid. But if the small grains are denser than the large grains the small grains will cascade through the interstices of a matrix of large grains and form either a basal layer or sometimes a concentrated layer within the matrix of large grains. Experimental results demonstrating some of

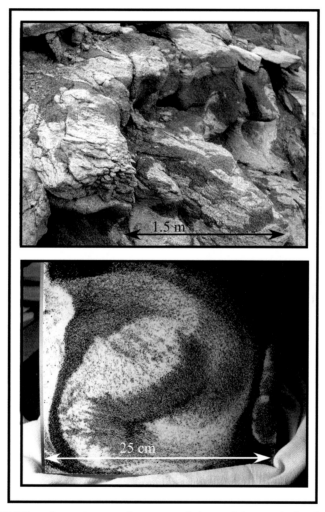

Figure 7.13 When the grains are of more equal size and the matrix is relatively dry of fluid (i.e., close packed solids), stirring promotes mushy patches. (upper) Field example at the Dais Intrusion (see upper right in map, Figure 7.1); (lower) laboratory experiment of dense packed solids of similar size in water having undergone shear by rotation of the container.

these effects are shown in Figure 7.12. Needless to say, similar processes are vitally important in many engineering operations and a highly useful summary is given by Aranson and Tsimring (2009; see also the review by Ottino & Khakhar, 2000). In summary, for bi- or multiply dispersive particulate systems (i.e., two or more particle sizes or types) the natural physical state is not a uniform mixture but an unmixed or segregated state arising from a wide spectrum of physical disturbances (Jaeger & Nagel, 1992). Any motion at all of granular slurries will bring about unmixing or sorting into layers, pods, and/or pipes. The tendency to segregate or unmix arises due to contrasts in particle size, density, shape, or resilience; it is an unavoidable process.

At the other extreme when the assemblage is relatively starved of fluid the granular mix moves more as a paste where, although there is some segregation, it is not nearly as complete as it is for more fluid rich systems. This type of motion or stirring of a crystal-rich mush, as remarked already, is also seen in the Dais in the upper southwest level in the trough area (see Figure 7.1) and is here shown, along with an accompanying experiment, by Figure 7.13.

Although far too numerous to detail here, the Dais Intrusion displays a wide spectrum of wonderful layering on a multitude of scales and, due to its unlimited exposure, it is a veritable laboratory available for research on layering processes.

# 8

# Compositional Characteristics of the Ferrar McDV Magmatic System

## 8.1 Introduction

As emphasized at the outset, process and sample context is everything to a firm understanding of magmatic systems. Volcanic and plutonic systems each have their own separate attributes; one has sound temporal context and the other sound spatial context, but rarely can the two be cogently connected in physics and chemistry. A great value of this magmatic system is that is has both; the spatial context of every sample as a record of the operational dynamics is known. That is, given the chemical composition of any specific sample, a clear reason can be recognized for this composition and its relation to the overall dynamics of development of the general system. In short, the system is ultramafic at the deepest levels, as represented by the Basement Sill and Dais Intrusion, and becomes increasingly differentiated, or less mafic, upwards at each level. Yet, remarkably enough, there is also a similar variation at each level within any single sill. Whereas the center of the Basement Sill at any specific location commonly reaches 20 wt.% MgO, the chilled margins and leading tip of the sill itself are always near 7 wt.% MgO. Moreover, in a radial direction outward from Bull Pass the average sill composition systematically decreases in MgO content. And there are some areas along the contacts, especially in the central part of the system near Bull Pass, where the granitic wall rock has become disaggregated and has noticeably contaminated the sill and these effects are clearly seen in the sill composition. Errant points on common chemical variation diagrams can thus be fully ascribed to a specific physical cause and, more importantly, chemical trends can be exactly mapped into spatial variations that can be related to dynamics.

This information from a system where the context is exactly known may prove invaluable as an interpretational aid when considering chemical information from systems (e.g., volcanoes, ocean ridges, and partially exposed subareal systems) where a similar level of context is not available. The plan here is to show the chemical variations at each level and upward through the system using various common projections, and to emphasize the correlations between chemistry and physical processes. This is particularly insightful and revealing knowing already from the foregoing what processes have been involved in producing these variations. In this way certain standard chemical variations can be understood in a much broader sense, and it is therefore essential here to consider these variations from a multitude of perspectives, both chemical and physical.

## 8.2  Full Suite Compositional Variations

Plots are given here of the variation in major element oxides and some trace elements for bulk rock samples as a function of magnesia (Figure 8.1) and silica (Figure 8.2). Although these variations, in and of themselves, are not highly instructive, they do clearly show the broad and continuous variations in composition spanning nearly 20 wt.% in MgO and 50–70 wt.% in silica. There are no gaps as are normally encountered in most volcanic suites covering this range of composition. This, of course, reflects the fact that this system has been spatially sampled in a particularly complete fashion, which is not possible in volcanic systems, where sampling is at the whim of the eruptive process itself. And it is further emphasized that in most petrologic studies it is data like this (along with trace elements etc.) that is all that is available, and subsequent petrologic modeling centers on reproducing these variations using various fractionation processes. Here the exact origin of these variations can be ascribed to specific well-documented processes, both chemical and physical.

On the magnesia plots, the variations in $TiO_2$, $Na_2O$, $K_2O$, and $P_2O_5$ mainly reflect increasing concentrations (with decreasing magnesia) of conserved or incompatible elements in the residual melt with loss of crystals. At the highest $TiO_2$ concentrations, which are rocks from the silicic segregations (open diamonds), the trend reverses with the appearance of Fe–Ti oxides in the crystallization sequence. The variations in CaO, iron, $SiO_2$, and $Al_2O_3$ across the full range of MgO, from 20 wt.% to about 7wt.%, mainly reflect the mechanical gain and loss of Opx crystals (see also Figure 1.1). At lower concentrations of MgO (i.e., $<\sim 7$ wt.%), these variations reflect the additional roles of plagioclase and augite. Although the trends are fairly coherent, there is some scatter (the star symbols), which is almost entirely represented by samples contaminated by granitic wall rock, as recognized in the field. This is also mainly true for the trace elements. Rb, Y, and Zr follow closely the variations in alkalis, $P_2O_5$, and $TiO_2$, increasing in the residual melts, while Ni and Cr are highly correlated with MgO or Opx content. The scatter of higher Cr values is not correlated with anything obvious in thin sections; it may reflect small inclusions in Opx. The scatter in Sr at

(a)

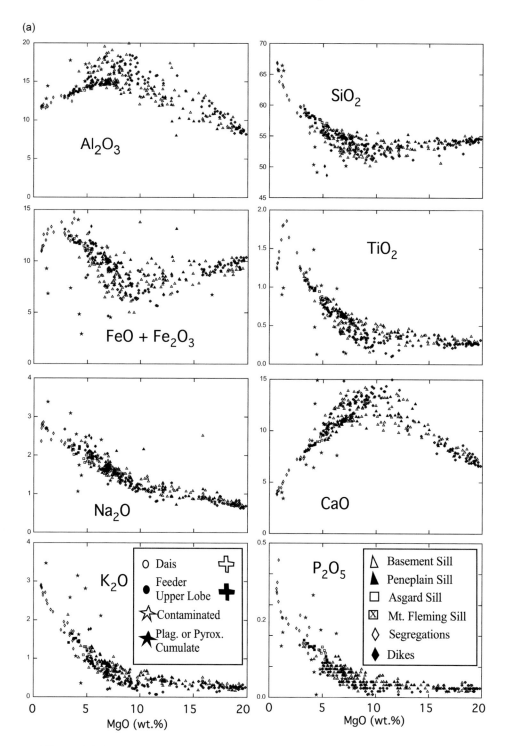

Figure 8.1 (a) Magnesia oxide variations for the full suite of Ferrar dolerites of the McMurdo Dry Valleys. (b) Trace element (ppm) variation with MgO (wt.%) for the full suite of Ferrar dolerites in the McMurdo Dry Valleys.

(b)

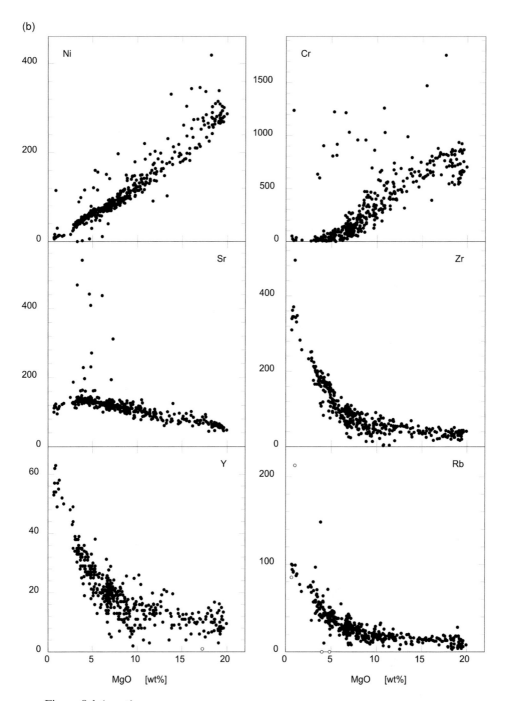

Figure 8.1 (*cont.*)

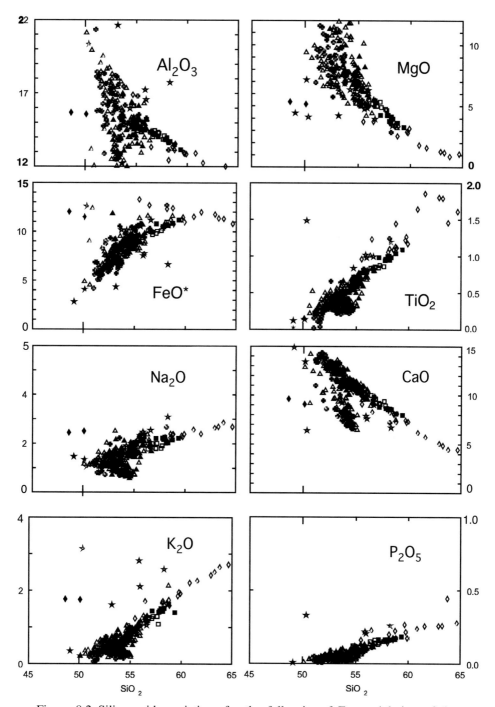

Figure 8.2 Silica oxide variations for the full suite of Ferrar dolerites of the McMurdo Dry Valleys

low MgO again reflects contamination by granitic melts near climbing contacts. Given that Ni and Cr will more strongly partition into olivine over pyroxene, their high concentrations here indicate that olivine was not an available or dominant phase in the present magmatic cycle.

In terms of dolerite classification, these are low titanium quartz tholeiites (LTQ), as opposed to HTQ rocks, and are considered depleted as they have less than ~1 wt.% titania at 6 wt.% MgO (e.g., Cummins et al., 1992). These compositional types (i.e., HTQ, LTQ, and olivine normative varieties) commonly form regional patterns as is found along eastern North America (Puffer & Philpotts, 1988; Ragland et al., 1992).

The oxide variations with silica (Figure 8.2) are not as revealing for this suite, primarily because the gain and loss of Opx, unlike for magnesia, has a silica content similar to the melt itself and does not strongly affect the bulk composition. The variations in the conserved elements ($TiO_2$, $Na_2O$, $K_2O$, and $P_2O_5$) are clear, but are not as distinctive, and the contaminated samples (stars) again standout, as do the two early dikes (filled diamonds).

Of the many possible discrimination diagrams used to classify similar rocks, conventional alkali–silica and AFM plots are useful (Figure 8.3). The non-alkalic and tholeiitic nature of this system is readily apparent; the contaminated samples standout, but there is no indication of the highly mafic nature of these rocks. On the AFM plot the ultramafic and strong compositional variation associated with sorting of Opx is readily apparent. There is a steady progression from the most mafic, highly concentrated Opx rocks of the Basement Sill to the Opx poor rocks of the chilled margins, the outer reaches of the Basement Sill, and the upper sills, ending with the least mafic or most fractionated rocks forming the silicic segregations (open diamonds). The contaminated rocks are again standouts (open stars). This sets the stage for the general chemical variations in this system. That is, all large chemical variations, at least for the major elements, can be traced to *spatial* variations relative to the sorting of Opx in the Opx Tongue, including vertical profiles across the Tongue or horizontally along the Tongue outward from Bull Pass.

This correlation between position, composition, and Opx crystal content is shown even more clearly in a plot of CaO against MgO (Figure 8.4). The Basement Sill (open triangles) and the Dais (open crosses) mainly define the overall trend down to magnesia contents of about 5 wt.%; the chilled margins and outer reaches of the Basement Sill having the lowest magnesia contents. The Opx-free sectors of the Peneplain Sill (filled triangles) and the upper sills, Asgard (open squares) and Mt. Fleming (filled squares), sequentially follow and lead to the silicic segregations (open diamonds) at the lowest magnesia contents. The break between Opx-controlled composition and equilibria due to augite and plagioclase

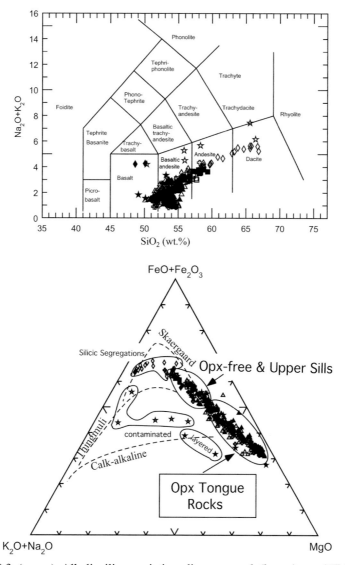

Figure 8.3 (upper) Alkali–silica variation diagram, and (lower) an AFM plot, indicating the non-alkalic, tholeiitic and highly mafic nature of some of these Ferrar dolerites

occurs near 7–9 wt.% MgO, as is often seen, for example, in Hawaiian lavas (e.g., Marsh, 2013).

A similar pattern is seen in a plot (Figure 8.5) of molecular components representing pyroxene (0.59 Mg+Fe) and plagioclase (2 Ca+Na), where the ferric and ferrous concentrations have been recalculated to an oxygen fugacity defined by QFM at 1,150°C. The upper trend is that due to the concentration of Opx

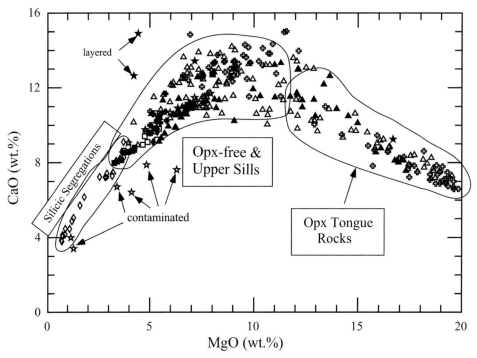

Figure 8.4 Bulk rock compositions plotted as CaO against MgO for the Ferrar dolerites of the McMurdo Dry Valleys (symbols are as given in Figure 8.1). There are three principal groups, as indicated: The Opx Tongue are the most mafic to the lower right, followed by rock without primocrystic Opx, such as chilled margins and upper sills, which is then followed by the silicic segregations. The plagioclase-rich layered rocks (dark stars) and the granite-contaminated rocks (open stars) show up off the main trend, which overall is much like that seen at Hawaii (e.g., Marsh, 2004).

(i.e., Opx control) and the lower trend is mainly due to the effects of plagioclase and clinopyroxene fractionation. The scatter in the main Opx trend, especially near the corner, is mainly due to the effects of sorting or layering in the Basement Sill, the Dais, and the Labyrinth section of the Peneplain Sill. This effect, which was also seen earlier in the Dais data, will later be singled out in more detail. These variations in crystal populations are also highly indicative of the overall mechanics of entrainment, transport, and emplacement of the Basement Sill, and these same variations give clear insight into the process of what is commonly mistaken as being due to in situ crystallization, fractionation, and differentiation.

## 8.3 Local and Regional Spatial Compositional Variations

It is common in the study of sills and some basaltic intrusions that only a piece is exposed or not even a complete vertical section through the body. Even less common is any indication of how, from what direction, and from where the body

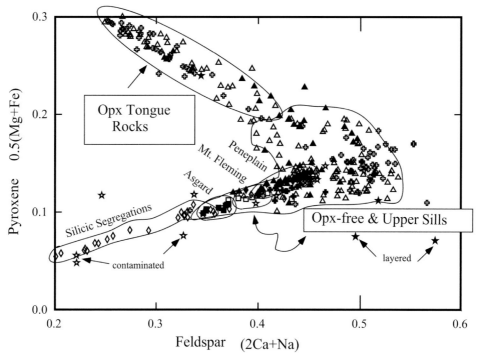

Figure 8.5 Variation of molecular components of "pyroxene" against "plagioclase" for the full Ferrar dolerite suite (symbols are as given in Figure 8.1). The Opx Tongue rocks dominate the pyroxene component progressively, followed by the Opx-free rocks, such as chilled margins and leading edges, and the least pyroxene control is for the upper sills and the silicic segregations.

was formed and filled. Here exists an unparalleled opportunity to characterize a massive individual sill in three dimensions, such that some general detailed physical and chemical insight or tools can be gained for understanding less well exposed bodies. With this intent, the presence of this extensive distribution of large Opx crystals in the Basement Sill, as mentioned previously, can be used as a tracer to follow sill emplacement and solidification. Using the compositional profiles determined from the extensive sampling throughout the Dry Valleys, the regional distribution of Opx is gauged using MgO as a proxy. To better understand the spatial context of the individual detailed chemical profiles proceeding outward from Bull Pass (Figure 5.6), these same profiles are presented on a regional topographic map (Figure 8.6). The distribution of Opx primocrysts appear regionally as a plume of material extending radially outward from Bull Pass, with the Opx Tongue becoming thinner outward in all directions. Starting north at Miller Glacier and going 100 km south to Cathedral Rocks, across from the Kukri Hills, the MgO content systematically diminishes outward from Bull Pass, as it also does in going east toward McMurdo Sound.

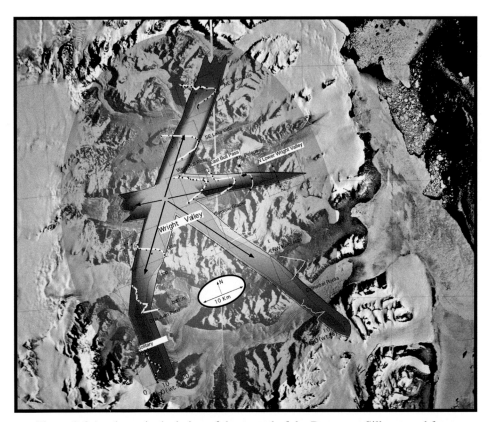

Figure 8.6 A schematic depiction of the spread of the Basement Sill outward from Bull Pass throughout the Dry Valleys, as indicated by the concentrations of MgO representing the Opx Tongue materials. The actual dynamics of emplacement was far more intricate and involved than this diagram might suggest, with, for example, multiple lobes pushing up Wright Valley along with southerly to Solitary Rocks as the Upper Lobe pushed through the Asgard Range to meet the Solitary Rocks lobe at the Kukri Hills.

That the Basement Sill (and Peneplain also) might continue much further south, as far as some 275 km south of Bull Pass, is suggested by two sills found at Roadend Nunatak, which is at the confluence of the Darwin and Touchdown Glaciers, near the Darwin Mountains and Brown Hills. Partial sections of two sills are exposed here, with the lower sill exhibiting an Opx concentration at about 50 m, containing up to 18 wt.% MgO (Kibler, 1981; as presented by Faure & Mensing, 2010, p. 430). The overlying Peneplain Sill is characteristically much more uniform, at 5–6 wt.% MgO. Given the overall style of emplacement, below and above the Kukri Peneplain, and partial thicknesses (each ~150 m) of these two sills, it is certainly tempting to link them directly with those in the Dry Valleys, but the much stronger MgO content over that found 200 km away at Cathedral Rocks

($<\sim$13 wt.%) makes this linkage tenuous. This occurrence may, instead, point to the existence of a separate, more local emplacement center in this region.

Given the strong spatial variations, both stratigraphically and longitudinally, it is important to appreciate in some detail the chemical variations exemplified by this distribution as indicators of the fundamental process of magmatic differentiation.

Starting in the center at West Bull Pass, chemical variations for the Basement Sill are shown by four plots (Figures 8.7 and 8.8). The samples from near the upper

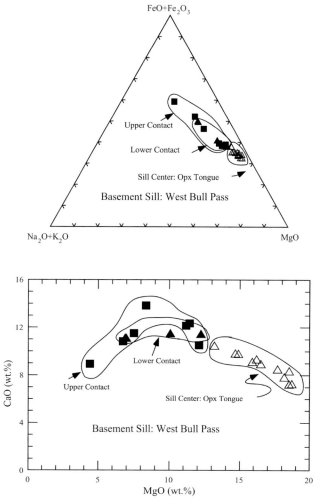

Figure 8.7 Chemical variation diagrams for the West Bull Pass section of the Basement Sill, illustrating the seminal chemical variations present locally, as also found for the whole Ferrar Dry Volleys suite. (upper) AFM diagram where the most fractionated rocks are those free of any Opx Tongue material. (lower) CaO vs MgO (wt.%) showing the large fractionation effect of the separation of the Opx Tongue minerals from the Basaltic Carrier Magma.

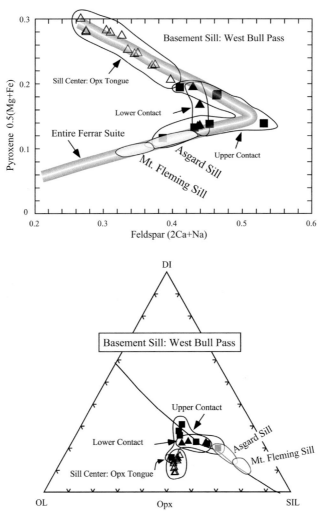

Figure 8.8 West Bull Pass section of the Basement Sill. (upper) Molecular proportions of pyroxene against plagioclase showing the strong control of Opx Tongue materials and the continuous transition into the more evolved primocryst-free compositions, which continue on with the Asgard and Mt. Fleming compositions. (lower) Normative basalt phase diagram at 1 bar projected from plagioclase showing the close adherence to the phase boundaries from the most MgO-rich rocks of the sill to the margins and continuing on to the most fractionated compositions of the Asgard and Mt. Fleming Sills.

and lower contacts are marked, along with those from the Opx Tongue or center of the sill. The span of compositions extend from about 20 to 4 wt.% MgO, which mainly reflects the concentration of Opx phenocrysts. The most evolved compositions are at both contacts, which is a feature found throughout the entire region; these near contact compositions are remarkably consistent over vast areas. The more strongly evolved upper contact composition reflects wall rock contamination accompanying the climbing contact here, which reflects ingestion of granitic wall rock and is specific to the central region near Bull Pass where prolonged heating produced extensive local re-melting of granitic wall rock. Elsewhere, outward from Bull Pass these effects are not found. Remarkably enough, the overall trend of compositions for this single vertical transect closely resembles the chemical variation for the whole system up to the level of the Asgard Sill (Figure 8.8). A comparison of these compositions with the compositional profiles at the most distal locations of Cathedral Rocks and Miller Glacier, each about 50 km south and north, respectively, of Bull Pass, shows a similar compositional spectrum (Figure 8.9).

The average compositions at Cathedral Rocks and Miller Glacier are each about 7.5 wt.% MgO. The Opx Tongue here is thin and the most magnesian compositions are about 13 wt.%. Even though these locations are about 100 km from each other, the overlap in individual compositions is remarkably similar. This reflects the physical process of sill emplacement and transport carrying fewer Opx primocrysts outward in the leading edge of the Basement Sill during emplacement. It is also to be emphasized that it is these distal regions that may represent the earliest emplacements. That is, this essentially same low MgO magma type gave rise to the upper sills, as can be judged by noting the compositional overlap in the normative basalt phase system between both the Asgard and Mt. Fleming sills with the Miller Glacier profile (Figure 8.9). More explicitly, the phase boundaries shown here are for the normative basaltic (MORB-like) system, and all the primocrysts-free compositions adhere closely to the low pressure (~1 atm) phase fields (see Chapter 14). Presumably the ultimate physico-chemical process of producing this "carrier magma," occurring laterally in the Basement Sill during emplacement, is similar dynamically and chemically to that giving rise to the upper sills and capping Kirkpatrick flood basalts (Figure 8.10). The Kirkpatrick data are from Thomas Fleming (personal communication, 2005; 2020; see also Elliot & Fleming, 2017, 2021).

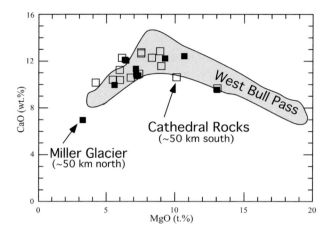

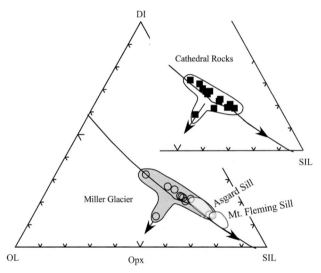

Figure 8.9 Chemical characteristics of Basement Sill sections about 50 km north and south of Bull Pass. (upper) CaO vs MgO and (lower) normative CMAS phase diagram projected from plagioclase showing close adherence to the experimental phase boundaries. These sections are about 100 km apart and, although the overall effects of Opx Tongue fractionation are still clear, the strength of the fractionation process is weaker.

## 8.4 Summary

The lessons learned from these spatial chemical variations are several (see Figure 8.11). First, if a piece of this sill was found and studied, the vertical distributions of crystals and compositions are not due simply to local cooling and solidification of an originally phenocryst-free magma. These large Opx crystals did not grow here and settle to fractionate the magma, even though at first thought it might seem so and is usually assumed so. Second, the differentiation that is seen

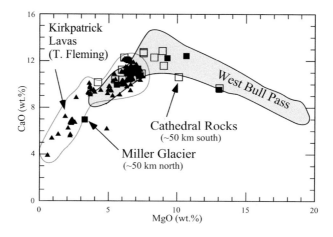

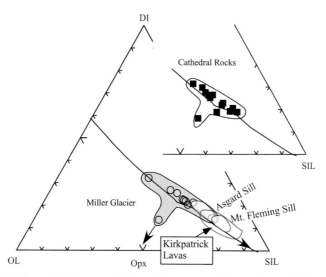

Figure 8.10 Chemical characteristics of Basement Sill sections at West Bull Pass and about 50 km north (Miller Glacier) and south (Cathedral Rocks) of Bull Pass as compared with the upper sills (Asgard and Mt. Fleming) and the regionally extensive Kirkpatrick flood basalts (latter data courtesy of T. Fleming). There is a complete compatibility between the Basement Sill compositions and a continuation from the Asgard-Mt. Fleming compositions to the Kirkpatrick Basalts.

locally over a vertical section is also exactly found laterally over the full extent of the sill, even down to local irregularities. Third, the compositions along the margins and at the leading fronts are exceedingly similar to the compositions of the Opx-free upper sills that contain no large primocrysts. It is this basaltic magma that dominates the system and is the driving fluid that runs the whole system. To be strict, the composition of this basalt does change systematically with time somewhat from the early lavas downward to the crystal-free magma of the margins

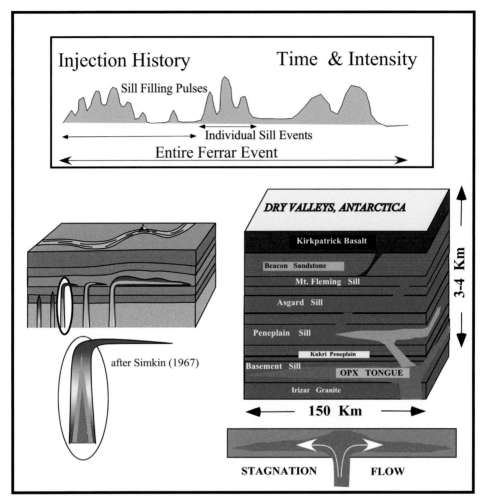

Figure 8.11 Schematic depiction of the overall temporal and spatial emplacement of the Ferrar Sills in the McMurdo Dry Valleys. (upper) The approximate or figurative history of emplacement; (lower left) the process of flow differentiation operating during ascent of the Basement Sill magma containing the massive slurry of Opx primocrysts; (lower right) and the overall distribution of the Opx slurry within the Basement and Peneplain Sills, along with an illustration of the concept of flow stagnation as the magma ascends and then spreads laterally to form a sill.

of the Basement Sill; part of this may be due to being bathed in large masses of Opx primocrysts (see Chapter 10). But because as previously argued the system developed from the top down, the basalt does become slightly more mafic with time. To critically identify this physical process of crystal entrainment, sorting, deposition, and its dramatic chemical effects, it is essential to understand something of the mechanics of entrainment and transport of the Opx crystals themselves.

# 9

# Crystal Entrainment and Transport

## 9.1 The Simkin Sequence

These large Opx crystals in every way play the same role here as do the olivine tramp crystals in Hawaiian lavas and lava lakes (Murata & Richter, 1966; Wright & Fiske, 1971). Hawaiian "Tramp Crystals" are large crystals of forsterite-rich olivine that are exotic to the basalt itself and have been entrained much as have these Opx primocrysts. These Opx crystals along with, strictly speaking, about 15 vol.% Cpx crystals reflect detailed mechanical processes deeper in the underlying magmatic mush column (Figure 1.1; Marsh, 2004). Sills and dikes containing similar axial concentrations of phenocrysts (mainly olivine) have long been recognized in Scottish sills and dikes (e.g., Drever & Johnston, 1967; Gibb & Henderson, 1996). This all builds on a fundamental concept introduced by Baragar (1960) bearing on his work in the Labrador Trough finding large concentrations of clusters of plagioclase phenocrysts concentrated in the center of sills. Baragar found that this sorting process also occurs in the transport of slurries in the paper and pulp industry. He initiated studies of this process among fellow scientists at the Geological Survey of Canada and suggested this might be a general process also found in magmatic flows; seminal fluid mechanical experiments realistically depicting this process were then performed by Bhattacharji (1967). Having seen similar distributions of olivine in Scottish sills on Skye, Simkin (1967) employed Baragar's concept of flow differentiation to suggest a sequence of ascent and emplacement as sills of crystal-laden magma. This Simkin Sequence is shown in Figure 8.11 along with the stratigraphic distribution of the Opx Tongue in the Ferrar sills (Marsh, 2004, 2015). To be more specific, in order to appreciate the delicateness of the Opx distribution in the Basement Sill, some background is required.

## 9.2 Particle Dynamics

A fundamental property of particles in ascending fluid is the tendency to sort themselves spatially by buoyancy and drag (i.e., size, shape, and density contrast) relative to the walls containing the shear flow. Heavy particles move away from walls and fall, relatively speaking, in the ascending fluid, establishing a central high concentration or tongue of particles. Although heavily investigated (e.g., Gibb, 1968; Komar, 1976), the detailed physics of this process is still not entirely clear, but migration is associated with small inertial effects in an otherwise essentially inertia-free flow. Because the crystal tongue falls progressively behind the leading tip of the ascending column of magma, the leading magma and the magma on the margins, having been stripped of crystals, is the most refined and is critically positioned to cool and solidify immediately after emplacement. That is, once ascent stalls and the magma begins to spread laterally, the leading crystal-free magma coats and chills, essentially cauterizing against the wall rock as it is progressively opened by the advancing magma. All the crystal-free magma about the sill margins is quickly chilled into fine-grained rock at the outer edges of the upper and lower Solidification Fronts. That is, at all contacts, major Solidification Fronts (Figure 9.1) begin to form and propagate inward at a highly predictable rate; the overall dynamics of Solidification Fronts has been presented in some detail in many earlier publications (e.g., Jaeger & Joplin, 1955; Marsh, 1989a, 1996, 2006, 2013, 2015); the specifics of crystal capture by advancing Solidification Fronts have been treated in some detail by Mangan and Marsh (1992).

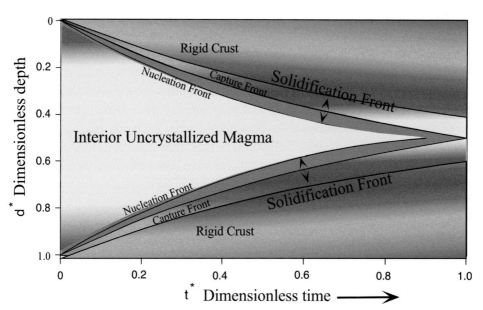

Figure 9.1 Schematic depiction of upper and lower Solidification Fronts migrating inward with time during cooling and crystallization of a sheet or sill of magma. The intensity of the colors is an indication of the prevailing temperature.

The upper Solidification Front captures any crystals existing near the contacts and only those at some depth away from the Solidification Fronts are able to settle fast enough to escape being overtaken and frozen into the Solidification Front, the rate of which decreases strongly as cooling slows with distance inward into the sill. The net effect for initially crystal-rich magmas is abundant crystals trapped near the upper contact and at sufficient depth a strong depletion as escaping crystals settle onto the lower, upward propagating basal Solidification Front, forming a dense collection of crystals. The overall distribution takes the form of a classic S-shaped modal concentration profile after solidification, which is found in many sills worldwide (e.g., Marsh at al., 1991; Marsh, 1996).

This Simkin Sequence of emplacement is beautifully revealed in the Basement Sill where the extensive Opx Tongue is found throughout the Dry Valleys. Although the locus of emplacement of sills is rarely found, here the distribution of the Opx Tongue gives direct insight into this filling point. As emphasized already, the Opx Tongue diminishes in thickness in all directions outward from the general area of central Bull Pass (Figure 8.6). From this area a series of petal-like lobes propagated outward from a central funnel-like magmatic feeder zone to establish the Basement Sill. The overall system appears to have developed from the top down with the tephras and local flood basalts first arriving from a local plexus of dikes emanating from low relief eruptive centers associated with high level sills breaching the surface (much more later). (There is a clear paucity of regional dike swarms.) Thickening of the basalts progressively capped the dikes forcing later arriving magma into more sills that intruded increasingly deeper in concert partially with the magma density, but also due to the overpressure in the system and loss thereof with venting. The Basement Sill intruded last in perhaps a slower, more sluggish fashion due to its high crystal content, but certainly under strong overpressure. Prior extensive heating of the country rock by earlier magma aided this slower, more deliberate ascent and lateral emplacement. That the emplacement process was quite likely pulsatile is indicated by internal irregularities in the MgO profiles. Stopping and restarting the emplacement process, as in volcanic reposes, causes telltale recording of the crystal distributions due to the steady, relentless advance of the upper and lower Solidification Fronts. Similar effects are apparent in the eruptive feeding of Kilauea Iki Lava Lake (Marsh, 2013, see figure 3). That is, every halt in the intrusion process allowed some settling of crystals, which was recorded by advancing solidification, as was also the restarting process. And upon resumption of the inflow, the new magma may have, in effect, "burned or wore back" some of the existing Solidification Front whether due to heat or increased drag or both. This pulsative process of emplacement is revealed in detail by subtle internal hot-chill zones within typical sills in this area as in the Beacon-Asgard Sill (Zieg & Marsh, 2012), and is very likely a general feature of most emplacement

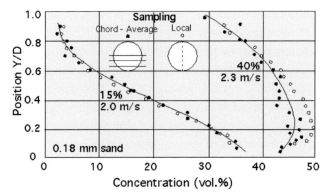

Figure 9.2 Concentration of particles in a cylindrical shear flow. Local and chord-average concentrations for 0.18-mm sand particles; notice especially that the lower particle concentration results in greater slumping toward the floor of the particle distribution profile, which tends to travel as a bedload along the floor, whereas at high concentrations the particles become more centralized in the flow (after Nasir-el-Din et al., 1987).

processes; but it is also particularly well recorded in the Opx distribution, as reflected by the MgO concentrations.

In particles being transported laterally in flow in a pipe or between parallel plates, without any effects of solidification, the distribution depends on the concentration of particles. At low concentrations (e.g., 10–15%) the particles can settle and be carried along as a bedload near the floor. As the concentration is increased up to that of maximum packing at ~50 vol.%, where there can be no flow at all due to rheological dilatancy, as mentioned in Chapter 7 (see also, Marsh, 1981, 2013), the particles will become increasingly evenly distributed across the flow. This is due to inter-granular pressure associated with particle drag that tends to make each particle equidistant from its nearest neighbors. This is shown here (Figure 9.2) in the experimental work of Nasr-el-Din et al. (1987) where, as the particle concentration is increased, the profile becomes increasingly parabolic, like the fluid flow itself.

## 9.3 Crystal Sorting with Solidification

When solidification is involved, the distribution of crystals is established by the shear flow, as they are progressively frozen into place as the Solidification Fronts steadily move inward from the upper and lower contacts, which is shown schematically by Figure 9.3; this is further exemplified experimentally in Section 9.4. If the flow is steady and does not stop, the entire concentration profile will be recorded in the progressive solidification. In magmatic systems this is not possible,

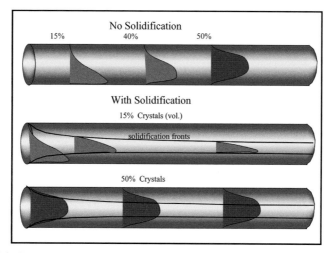

Figure 9.3 Schematic illustration of the effect of particle concentration on the distribution of particles in a cylindrical shear flow without (upper) and with (lower) the effects of solidification on the retention or recording of the particle distribution during flow, as in the Basement Sill. A flow that stops, allowing solidification to advance inward, and then restarts will concentrate the particles in the center of the sill. This is intended to portray the distribution of Opx crystals in the Basement Sill as it extends outward from its filling or stagnation point in Bull Pass.

as the flow must stop somewhere in a finite time, but for fairly thin sills and dikes a rather complete distribution may be captured. What is more likely is that the flow either starts and stops or slows and speeds up, off and on, leading to pulsative emplacement, and these hiatuses will also be recorded in the Solidification Fronts, which is shown schematically by Figure 9.4 (Charrier & Marsh, 2004, 2005, 2021). As can be easily imagined, the time series of possible combinations of starts, stops, slowing, and speeding up for magmatic emplacements is certainly vast, not to mention the additional effects of unusual sill geometries, secondary nucleation and growth, and clustering effects. This rich subject has been explored in some detail by Charrier (2010), where she shows how nearly any profile of crystal concentration can be related to the dynamics of emplacement. This method has been applied to the Basement Sill, and a typical modeled profile is shown by Figure 9.5 (Charrier, 2010; Charrier & Marsh, 2021). It is particularly noteworthy that this shows two profiles through the Basement Sill in the broad lobe going north toward Lake Vida, but at much different parts of the lobe miles apart. The crystal distributions, nevertheless, are broadly similar, reflecting a similar history of magma movement. The shading is used to illustrate four possible episodes of filling, each subsequent episode causing some stripping and back stepping of the distribution of crystals as the inflow stops and restarts; similar analysis has been done by Charrier (2010) for other lobes of the Basement Sill.

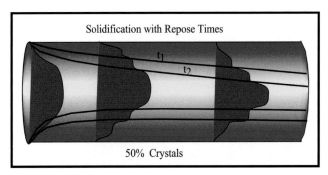

Figure 9.4 Schematic illustration of the effect of solidification and repose times during magma emplacement as a shear flow on the final distribution of crystals when the initial concentration is ~50 vol.%. During periods of repose the solidification fronts, both upper and lower, continue to advance, freezing in the concentration of crystals such that, when the flow resumes motion, steps or variations in crystal content are preserved in the final rock record.

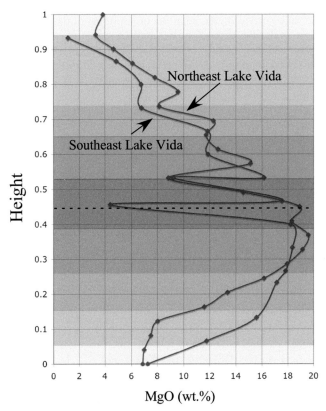

Figure 9.5 An illustration of the similarity in concentration of Opx crystals (as inferred by MgO (wt.%) content) in the Basement Sill for two profiles ~10 km apart near Lake Vida in Victoria Valley. The broad symmetry in the profiles may indicate repose times during sill magma emplacement, which repose times are indicated by the various levels of shading, reflecting four periods of activity.

## 9.4 Experiments on Mush Column and Sill Primocryst Dynamics

Once the overall geometry and sequence of development of the McDV sill system was basically understood, attempts were made to simulate experimentally its operational dynamic behavior. That is, the main idea was to build a system that matched, more or less, the true system using numerical methods and actual tanks of fluid in order to, in essence, *restart the system* and study how it may have operated when it was functioning some 180 Ma ago. Each approach at this early stage is, clearly, of a purely preliminary and reconnaissance nature, but the results are, nevertheless, enlightening. And, although initially intended to provide insight into a series of questions, in the very performance of these experiments a much better understanding was achieved of how particle-laden "magma" might act within a mush column of sills; thus, leading to making better observations in the field.

A series of fundamental questions presented themselves, and these questions guided this experimentation. Namely:

Once a sill is established, does later magma coursing upward through the system easily re-enter the earlier sills and lead to clear multiple injections?

Does magma enter a sill only when it is allowed to grow laterally?

Does the system nucleate new sills only when the upper sills become capped or solidified?

As the system develops, how seriously is the crustal thermal regime affected over the life of the system?

How must the adjoining feeding zones or conduits between sills be placed to allow the continuous transport of large crystals (particles) upward through the entire system?

Is there a critical plumbing geometry whereby large crystals cannot reach the surface?

How does ongoing solidification affect the ability of the magma to transport crystals upward through the system?

What styles of flow through the sills allows for well layered deposits to form?

Three series of experiments, numerical and physical, were conducted to better understand these questions, centering on the intrinsic nature of the inter-connectedness of the crystal-laden magmatic flows making up this system. To first gain a basic understanding for the fluid flow through the system a numerical model was used to visualize the flow and evaluate its complementary thermal effect. A second set of physical experiments involve simulating the transport and capture of particles in a solidifying fluid (hot wax) forming sills in a viscoelastic (gelatin) crust. The third set of experiments involved building an actual series or stack of interconnected plastic tanks, representing sills, to evaluate the distribution of particles as related to the larger dynamics of a "magmatic slurry" coursing through a magmatic mush column within the crust.

### 9.4.1 Numerical Fluid and Heat Flow

Starting with a basic geometric form representing three simple sills, each 350 m thick stacked in a crust of about 4 km thick and 50 km wide, the fluid flow was first calculated for magma entering the bottom and flowing upward to the surface (see Figure 9.6). (The calculation used a Finite Element Modeling program (COMSOL Multiphysics Package).) The two-dimensional flow is viscous, isothermal, and is allowed to continue to flow upward throughout the system and "erupt" onto the upper surface. The hotness of the colors indicates the magnitude of the flow, and the small arrows indicate flow direction. Once a sill is established, even though it may still be molten, magma flowing through to the surface will not penetrate and re-circulate into existing sills. The lowest sill here is in the process of being emplaced and thus shows lateral flow. Apparently only if the system becomes capped will there be sufficient internal over-pressure to extend an existing sill or initiate a new sill. The path of the fluid to the surface is direct, moving at a continuous but lesser flow rate from the lowermost to the upper sills.

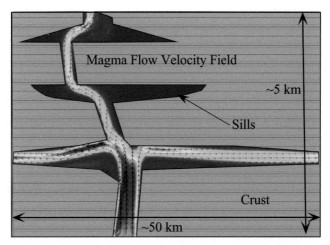

Figure 9.6 A simple numerical calculation illustrating the probable flow of magma upward through a stack of sills embedded in the upper crust. The two-dimensional flow is viscous, isothermal, and is allowed to continue to flow upward throughout the system and "erupt" onto the upper surface. The hotness of the colors indicates the magnitude of the flow, and the small arrows indicate flow direction. Only when a sill is developing or growing does magma enter the sill, as in the lowest sill.

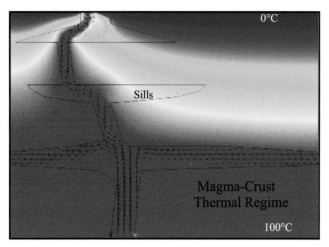

Figure 9.7 A numerical calculation demonstrating the heating of the crust by the presence of an active magmatic system, as in Figure 9.6, which here has been operating for ~0.5 Ma. There is ample opportunity for extensive melting in the crust adjoining the magmatic system.

Using the same system, the long-term thermal effect on the adjoining crust was evaluated (Figure 9.7). The system has been running for ~0.5 Ma and the initial crust thermal regime is indicated by the color intensity with 0°C at the surface and 100°C at the base. The magma has an initial temperature of 1,200°C. The presence of these hot sills has a profound effect on the thermal regime of the crust and depends on the sequence of sill emplacement, sill volume, and especially the lateral extent. This simple example also suggests that an extensive hydrothermal system is certain to be induced in the adjoining crust, especially when the sills are in a thick section of the sediments like the Beacon Supergroup that was covered by a shallow sea. And, consequently, the earlier formed sills will be subject to ongoing hydrothermal alteration driven by this enhanced heating of the crust due to the presence of the mush column. The pervasive high temperatures surrounding the sill system are in the range of 900–1,000°C.

The actual system most certainly did not run continuously for 0.5 Ma but, as inferred from the field relations, magma flow was periodic or pulsatile. The thermal effect on the adjoining crust is cumulative, and to give some perspective of this the conductive cooling of a sill-like stagnant sheet is shown as Figure 9.8. Here the sheet is of thickness 2L, time (t) is given in dimensionless units of $Kt/L^2$, where K is thermal diffusivity ($\sim 10^{-6}$ m$^2$/sec), and the dimensionless temperature (top) is scaled by the initial magma temperature, $T_o$, relative to the initial wall rock temperature, $T_w$. This history of cooling is also identical to that of a dike. Notice especially that the contact temperature is always at or below the average of the two

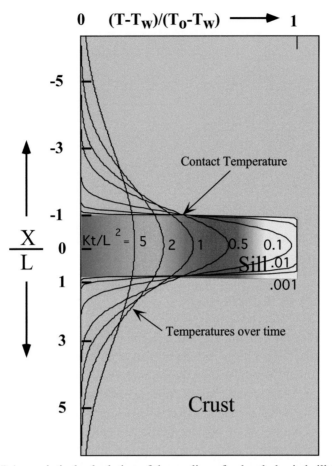

Figure 9.8 An analytical calculation of the cooling of a deeply buried sill (or dike) of arbitrary thickness and initial temperature. There is a substantial amount of heating of the wall rock, which becomes increasingly severe with repeated emplacements of magma.

(i.e., magma and wall rock) initial temperatures; this example does not include the effect of latent heat, which increases the contact temperature slightly from 50 percent to 65 percent of this sum (see Wright & Marsh, 2016). More important to appreciate is the wide thermal anomaly left in the adjoining crust after solidification has completed, such that ensuing emplacements will likewise cumulatively increase the crustal temperature with contact temperatures being progressively higher. This effect is, of course, much more enhanced when the magma is not stagnant, but continues to flow through the crust for any significant period of time. Hence, the widespread melting of the granitic wall rock in Bull Pass is a direct reflection of the presence of a locus of magma flow feeding the entire system and severely heating the country rock.

### 9.4.2 Hot Slurry Transport with Solidification during Dike and Sill Formation

#### 9.4.2.1 Particle Flows in an Open Channel

As a precursor to more detailed and well controlled experiments, a series of simple exploratory experiments were conducted to gauge the effects of solidification on molten wax (i.e., "magma") carrying particles. Molten wax laden with glitter to form a well-mixed slurry was poured in an open, inclined cylindrical trough to simulate a propagating sill or lava flow. As the wax progressed down the trough, solidification fronts migrated inward from all boundaries, including the open top (Figure 9.9), and the particles commenced to settle, forming a bedload deposit that was progressively captured by the inward solidification. The lateral edges and leading front of the fluid were always free of particles and froze free of particles until the later stages of cooling. To simulate the process of reinjection, pouring was stopped prior to final solidification and then recommenced through a hole at the upper end. In each instance the new hot fluid inflated the earlier solidified, but still easily deformable, envelope of wax, forming a breakout, usually near the toe of the body. The repose time between pours was a fraction of the full solidification time, which was about three or four minutes, during which time all the particles settled on the lower or floor solidification front, forming a distinct layer. Each reinjection was marked by a separate layer of sediment and the distance between each deposit closely reflects the progress of solidification during the repose times (Figure 9.9).

These simple results showed that broadly similar layers of Opx in the Basement and Peneplain sill might be due to such a process as already discussed. In these experiments the low viscosity of the wax and the slightly too heavy glitter allowed for rapid settling and formation of a strong basal layer. If the slurry had been more concentrated, the distribution of particles would have more been diffuse and not so strongly concentrated on the floor. But these results do demonstrate the characteristic effect of solidification and particle transport and settling, leading to distinctive distributions of crystals in sills. These results are also perhaps more akin to the dynamics of lava flows and, to emulate similar effects for a propagating sill or dike, rather than a flow in an open channel, the flow should propagate into and through a suitable "continental crust," which is what the next series of experiments were designed to do.

#### 9.4.2.2 Propagating Dikes and Sills

The theoretical and experimental aspects of the mechanics of dike propagation and sill formation have been well studied (e.g., Fiske & Jackson, 1972; Hyndman & Alt, 1987; Lister, 1991; Koyaguchi & Takada, 1994; Takada, 1994; Hirata, 1998; Dahm, 2000; Watanbe et al., 2002; Roper & Lister, 2005; Galland et al., 2006; Kavanagh et al., 2006; Rivalta & Dahm, 2006; Ablay et al., 2008; Menard, 2008;

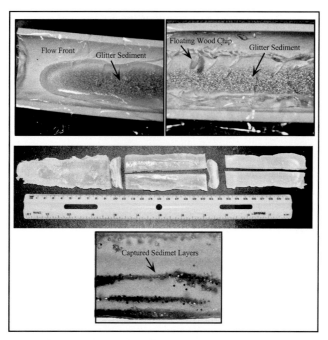

Figure 9.9 Experiments simulating flowing magma (molten wax) containing entrained crystals (glitter). A well-mixed slurry of molten wax and glitter, poured into an inclined open half cylinder, flows downward, undergoing progressive solidification. (upper right) A chip of floating wood shows the depth of the flow, and the glitter sediment forms a bedload in the center of the flow. (upper left) At the leading edge of the flow the sediment lags behind and has migrated away from the walls. (center) A typical end result sliced lengthways and cross stream reveals the distribution of particles in the solidified wax. (bottom) Starting and stopping the pouring, as in multiple injections in a sill or lava, allows some solidification between each event, which produces several sediment deposits in the final solid material.

among others). These experimental works have considered the 2-D and 3-D forms of dikes, the rate (steady, non-steady) of propagation in viscoelastic media as a function of the dike fluid rheology, the role of stress (and topography) in the "crustal" medium on steering dikes, the transition from dikes to sills by neutral buoyancy, the effects of multiple dikes on the stress field and subsequent dike orientation, and the effects of rigid boundaries in sill and dike formation, among many other studies. By and large, an upward propagating dike reaching a point of neutral buoyancy or encountering a change in the prevailing stress field, due to topography or crustal structure, among other factors, encourages sill formation (e.g., Ryan, 1993).

In these experiments the dike and sill fluids used are generally water or glycerin, and sometimes air. Few, if any (but see Currier & Marsh, 2015) experiments have considered the interplay of the effects of solidification and crystal content on the form of the subsequent intrusion and the transition of dikes and sills from one to another in the same event and, especially, how the final distribution of transported crystals reflects the dynamics of emplacement and filling. The clear role of the massive Opx Tongue and solidification may well have been a major influence on emplacement of the Basement Sill and Peneplain Sill. An understanding of the interplay of these effects, that is, solidification and crystal content, may furnish additional insight into the overall development of this magmatic system.

In this regard, a large series of experiments were thus performed using particle-laden molten paraffin as a proxy for crystal-rich magma to learn about the formation of a magmatic mush column, including the effects of particle sorting and the transition from dikes to sills to plutons (e.g., Currier et al., 2010; Currier, 2011; Currier & Marsh, 2015). As in many other studies, gelatin was used as an injection medium, which plays the role of a viscoelastic continental crust. The stiffness of this medium can be modulated in many ways by, for example, varying the portions of water versus gelatin, mixing in vegetable oil, adding tiny solid particles, and by temperature. The clarity of gelatin allows the ongoing process of injection to be easily visualized and in the end the gelatin can be washed away to reveal the form of the solidified intrusion. For magma, molten paraffin laced with ultra-fine particles (glitter) is used so that the particle-based Reynolds number (governing Stokes settling) matches that of the true magmatic system. And, to make the solidification time competitive with the rates of particle settling, as in the real case, paraffins of various higher molecular weights were used to get the right melting point. Large clear plastic tubs of gelatin simulate the crust, and, through a hole burned in the base, the "magma" is injected via a large, insulated syringe.

After developing some facility with the technique, the first observations found that with air at the head of the incoming fluid, dikes always form, but with no air, sills form immediately. (Magma ascending into the crust will almost always exsolve a vapor phase, leading to dike transport.) To remove the effect of the tub bottom as a rigid boundary a small pipe is inserted up from the floor, raising the injection point. The principal results mimic to a remarkable degree the features seen in the McDV sill complex, and especially the Basement Sill and the Opx Tongue (see Figure 9.10). That is, from an initial dike a sill forms, the position of which can be controlled by placing a tissue, or any other of the very slightest material disturbance, in the gelatin. The initial "magma" is particle free, but the plume of particles is entrained and follows, forming an extensive Particle Tongue in the sill and also in the dike. And, as in real dikes, the particles are restricted to the center of the dike and sill, encased in true chilled margins. Even though the injection process is smooth and steady, the sills and dikes always propagate in a jerky fashion, which is clearly recorded by "growth rings" in the wax. This

Figure 9.10 Sill emplacement from an ascending dike modeled with wax laden with particles as magma penetrating a crust of gelatin (left). The darkness indicates the concentration of particles entrained in the wax, and the tendency of the particles to migrate to the center of the dike is shown in the cross-section in the upper right. The growth rings indicating the stop and start nature of the injection process are shown on the lower right. A possible similar effect is sometimes found in the chilled margins of the Ferrar sills, where a sample of this feature is also shown.

apparently reflects the quenching and restarting at the crack tip (Currier & Marsh, 2015). We have long noted similar delicate features in the dolerite sill contacts, and these may also be a measure of time during injection (Figure 9.10). Evidently, the magma suddenly advances locally, is quenched, builds up pressure, advances again, quenches, and repeats the cycle over and over (Currier & Marsh, 2015; see also Figure 9.11). If the load of "magma" is large and injected slowly and unsteadily, mimicking a pulsatile process as envisioned for the McDV system, the resulting structure mimics a small mush column (Figure 9.11). Small, bulbous apotheoses form but solidify too quickly to become larger successful bodies, and the crystal plume extends into the initial sill and then continues upward in the dike but does not reach the upper sill; this may be similar to the early development of the McDV system. Below a certain level of gelatin strength, large bulbous, pluton like masses form, rather than large sills. It proved elusive to develop a system where the magma broke through to the surface and under the right conditions solidified to cap the system, forcing further sill development. To study this process a large-scale system of interconnected tanks was built (see Section 9.4.3).

These experiments revealed a remarkable controlling role of solidification fronts bordering all magmas in regulating or throttling the process of sill or laccolith

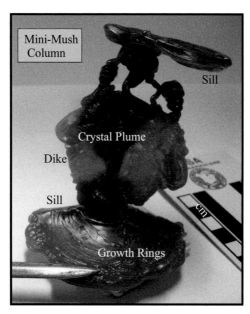

Figure 9.11 A miniature mush column made by injecting wax, laden with particles, in a pulsatile fashion into a weak gelatin crust. The initial sill is arrested, and the wax magma continues upward forming a bulbous irregular body that is also arrested and then emits several "vents," leading to a higher-level sill. The plume-like concentration of particles is indicated by the darker areas of the wax.

formation. This series of carefully scaled experiments used two contrasting magma analogs, water and molten wax, that are again injected into viscoelastic gelatin representing continental crust. The main results of these experiments are quoted here from Currier and Marsh (2015, p. 211):

In non-solidifying, water-style experiments, intrusion is a relatively simple process. In contrast, wax magma emplacement displays a vast palette of compelling behaviors: propagation can slow, stop, and reactivate, and the directionality of lateral growth becomes much more variable. Small flow deviations in water-based intrusions are likely the product of flow instabilities, the result of injecting a viscous fluid along an interface (similar to Hele-Shaw cell experiments). However, the much more complex emplacement style of the wax experiments is attributed to solidification at the leading edge of the crack. The overall effects of solidification during emplacement can be described by a non-dimensional parameter measuring the relative competition between the rates of crack propagation and solidification at the crack leading edge. In this context, laccolith growth mechanics can be separated into three distinctive characteristic stages. Namely, I: A thin pancake style sill initially emanating radially from a central feeder zone, II: As solidification stalls magma propagation at the leading edges, enhanced thickening begins, forming a true, low aspect ratio laccolith, and III: As stresses accumulate, tears and disruptions readily occur in the solidified margin causing fresh breakouts, thus reactivating lateral growth into new lobes. The competitive combination of these latter stages often leads to a characteristic pulsatile growth. The unexpected richness of these results promises to add fundamentally to the basic understanding of laccolith growth mechanisms, and also adds key observations to growth processes to be sought in field studies that have been hitherto suspected but rarely observed. Despite seemingly simple shapes, the growth history of laccolithic intrusions is likely quite complex.

The basic style or overall nature of these effects can be seen in the field in the long, sweeping, and delicate propagating edge of the Basement Sill in McKelvey Valley, directly north of Bull Pass.

### 9.4.3 An Experimental Magmatic Mush Column

To gain some understanding of the interplay between particle or crystal transport throughout a mush column relative to sill placement, fluid flux rate, and the placement of feeder zones or conduits between sills, a large-scale experimental apparatus was built (Flanagan-Brown & Marsh, 2001). In addition to the questions posed earlier in this section, perhaps a more immediate question concerns the controls on determining how crystals are delivered to and through various parts of the system; especially, what combination of parameters prevents crystals from reaching the upper sills and erupting onto the surface. This is a rich subject with much broader similar but distinct seminal work (e.g., Moritomi et al., 1982, 1986, among many others).

At any instant at various locations the true system contains fractionated and primitive melts as pools of nearly crystal-free magma, pools of crystal-rich magma, thick beds of cumulates, open conduits, and conduits congested by cognate and wall rock debris. All boundaries of the system are sheathed by active solidification fronts. With the wide range of local characteristic length scales, there is a commensurate range of solidification time scales. This creates a complicated series of resistances to magma flow and provides a variety of distinct local physical environments for the chemical modification of the governing magma. The system is driven by over-pressure and buoyancy from the addition of new melt from below. The over-pressure propagates upward by moving magma, which flushes conduits, disrupts cumulate beds, and pools or purges sills. As repeatedly emphasized, a critical aspect of this process is the entrainment, transport, and deposition of crystals throughout the system. Like the Opx Tongue material found here in the McDV system, elsewhere picritic lavas charged with entrained (tramp) olivine of a wide compositional range are erupted at many systems (Jan Mayen, Kilauea, Reunion, etc.) as a clear expression of this process. The size and abundance of these crystals is correlated with eruptive flux (Murata & Richter, 1966), suggesting this to be an important indicator of the overall dynamics of the mush column.

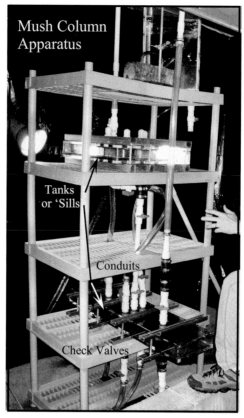

Figure 9.12 The experimental setup of the series of tanks (sills) interconnected by a series of conduits containing shut off and check valves, allowing particle laden fluid (water) to be pumped through the system, from the base, in an arbitrary configuration. The scale of the system is indicated by the hand and foot in the picture.

An attempt was made to model some of the basic features of this process by studying the entrainment, transport, and deposition of particles in a vertical stack of sills (Plexiglas tanks) connected by resistive conduits (check valves), over-pressured from the base, and open at the top (Figure 9.12). The system is about two meters in height with water as the "magmatic" fluid and particles (wax and synthetic polymers) with Reynolds numbers closely approximating actual crystals.

A typical setup and experimental run are first described, followed by observations and discussion. The system was first filled with clear water and all bubbles purged. Particles of a range of sizes and similar, but distinct, densities making up a dense slurry are placed in a basal reservoir and also in the lower input pipe to be carried upward by the fluid pumped from the base. The base state is a sustained over-pressure that permits a steady effusive flux of particle-laden fluid throughout the system. To establish this state, the main conduit resistance (check valve tension) must first be overcome. As the critical moment of flow is approached, the system commonly exudes a high-pitched vibration emanating from the check valves, which may be akin to the well-known harmonic tremor of active volcanic systems. A time-series of over-pressure pulses was also sometimes employed, designed to mimic eruptive episodes of volcanism. Particle deposition occurs on the sill floors, forming beds around the conduits, and cumulate piles are repeatedly disrupted by flux associated with over-pressure pulses and flushing of choked conduits. The largest particle concentrations are in the basal sill and particle concentration decreases upward in the system. The magnitude and frequency of over-pressure pulses controls the ability of the system to erupt particles. Cumulate deposition is restricted to the conduit areas in each sill unless the flow is made to spread throughout the sill by horizontally separating the inflow and outflow conduits. Spreading flows, which form laterally graded cumulates, are also achieved by simulating outward lateral propagation of the sill via sidewall taps allowing fluid loss. The overall dynamic behavior of the system suggests that the time history of the cumulate eruptive flux, as measured by crystal size, abundance, and composition, may carry information critical in evaluating the deeper plumbing and dynamics of similar mush column magmatic–volcanic systems. Next are some more detailed observations, where the schematic setup is convenient in these discussions and explanations (Figure 9.13).

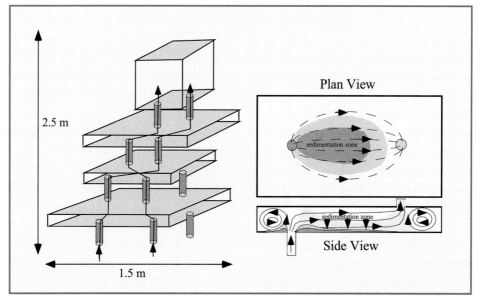

Figure 9.13 A schematic depiction of the experimental apparatus in Figure 9.12, illustrating the approximate size of the system, the interconnecting conduits, and the possible courses of fluid flow through the system. The plan and side views indicate the general pattern of fluid flow in transiting from one conduit to another along with the likely zones of sedimentation and particle sorting.

### 9.4.3.1 Harmonic Tremor

The first check valves used were of the ball and spring variety. As the overpressure was increased to the point of overcoming the check valves, the entire system suddenly began to vibrate and emanate a distinct high-pitched sound, which increased in magnitude until the check valves opened allowing fluid flow. In some fundamental respects, this setup is similar to Hill's (1977) model of a swarm of interconnected fluid filled fractures or dikes connected to one or more wing fractures where resonance increases with pressure in the conduit. causing the shear fractures to fail seismically at fixed time delays. This is the explanation by Stroujkova and Malin (2001) of harmonic tremor recorded at Long Valley. In the real situation the choking of conduits by blocks from the wall rock would seem in many respects to function as a suitable check valve. A general explanation of harmonic tremor may be much more involved (e.g., Hellweg, 2000; White & McCausland, 2019), but this experimental finding may illustrate a critical feature of mush column operation.

### 9.4.3.2 Conduit Spacing and Particle Trapping and Sorting

As repeatedly emphasized already, a central feature of slurry-based systems involves the mechanics of transport, sorting, and deposition of particles as a function of conduit size and spacing and fluid flux. At major volcanic centers the abundance and size of olivine crystals carried by the magma are found to be proportional to the magnitude of the eruptive flux. The ability of the ascending magma to entrain and transport significant amounts of crystals may depend intimately on specific aspects of the plumbing system. For example, if the lateral spacing between successive conduits is too large, in transiting from one conduit to the next the horizontally traveling magma steadily loses crystals to settling and may be barren of crystals by the time it reaches the exit vent (see Figure 9.13). There is, thus, a critical relation between the abundance and size of crystals and the distance between the entering and exiting conduits in any sheet-like chamber. This relationship depends on the rate of lateral flow and the settling rates of the entrained crystals.

That is, if the time for settling is less than the time for the magma to make the transit, the crystals will be lost from the flow. But because there is always a spectrum of crystal sizes in the flow, it is this spectrum that will be modified. Also, in consideration of conservation of mass in the flow, magma entering a chamber or sill will experience a major decrease in mean velocity relative to that in the conduit. This decrease in velocity is in direct proportion to the ratio of the cross-sectional areas of each flow region (i.e., pipe versus sheet-like rectangle). Magma entering and traversing a sill has a greatly reduced capacity to carry phenocrysts, whereas magma entering a conduit has an enhanced capacity for entrainment of crystals and, in reality, erosion of the conduit itself. Considering that the ensemble of entrained crystals may be of a wide range in composition and size, conduit spacing may thus be an effective filter to sort crystals and, in some cases, virtually eliminate them from the ascending magma. Moreover, the crystals settling from the flow will form a size and compositionally graded layered sequence.

In this regard, that the Basement Sill magma carried large crystals out to large distances indicates a strong, slurry-laden flow. The Opx load in the overlying Peneplain Sill is much more local, emanating from near the apparent central vent or conduit in Bull Pass, and travels outward as far as the Labyrinth, where the crystals become increasingly smaller, but apparently not to greater distances. How the Asgard and Mt. Fleming sills were fed is not yet clear, but there is no sign of the Opx load in either of them, which may simply indicate that the early eruptive flux was too weak to carry such a slurry.

### 9.4.3.3 Vent Base Surge Deposits and Far Field Eddies

In the flow field sketched in Figure 9.13, as the flow enters the sill and spreads and moves laterally to the exit conduit, it also drives a single convective cell or eddy in the stagnant or far field part of the chamber. Immediately about the vent a conical mound of cumulates forms, which is highly asymmetrical, being distended toward the other active vent. This mound tends to over steepen with time and cascade outward in all directions. This causes a distinct sorting into well-formed layers in response to the differing angles of repose for particles of different shape and/or density (e.g., Jaeger & Nagel, 1992; Jaeger et al., 1996; Makse et al., 1997; Aranson & Tsimring, 2009). This process is greatly augmented by time variations in the up flow from the vent, as is typical in long-term volcanic eruptions, and it leads to, in effect, a delicately sorted, low Reynolds number base surge deposit (Figure 9.14). The very small particles are kept suspended in the flow and most are carried laterally and into the exit vent, but those that become caught up in the cell trapped in the stagnant end circulate extensively and eventually form a well sorted deposit of unusually fine particles. The net result around the vent is to produce a layered deposit that is strongly asymmetrical in thickness, areal extent, and particle size. It is this style of process that may have contributed to the delicate sorting in the layered deposits of the Dais Intrusion. The lateral flow from the vent in Bull Pass to the ponding area at the Dais supplied a variety of crystal sizes and abundances, along with apparently significant volumes of crystal-free magma allowing, for a variety of these sorting processes to take place.

### 9.4.3.4 Flushing of Choked Conduits

Sorting and deposition around vents, especially in regard to the form and size of the vent is readily apparent in these experiments (Figure 9.14). Funnel shapes and

Figure 9.14 A typical base surge-type distribution of particles deposited on the floor of a "sill" near a conduit where the flow enters from below. The sorting of particles is indicated by the stronger concentration of light-colored particles in a ring-like distribution about the conduit.

vents made principally of the deposited particulate material in concert with the time variation of the flow are also controlling factors in building layering around the vents (Figure 9.15). That is, as an episode of transport wanes, particles begin to collect in abundance near the inlet, even to the point of filling the conduit itself as the event ends. The conduit, which contains a check valve, thus becomes choked with particles. When the next transport phase begins, if the flow is sudden and strong, the particles are erupted from the vent as a concentrated mass or slug, which develops into a gravity current that forms a distinct disconformity within the cumulate sequence. If the flow intensity ramps up more slowly, a dense slurry of particles is formed in and near the vent, which progressively partitions the smaller and lighter particles into the flow, leaving the larger and denser particles near and in the vent itself (Figures 9.14 and 9.16). At the close of the flow episode, close inspection of the deposit in the vent itself shows a steeply-dipping, well-sorted layered sequence. Deposits very much akin to these are exhibited in the coarse-grained body in central Bull Pass (Figure 6.9), and a more detailed spatial study of

Figure 9.15 The layering of particles as the fluid transits horizontally between conduits, progressively filling the "sill" and sorting the particles. Much of the variations in layering comes from variations in the fluid flux over time.

the crystal sizes and modes may give critical insight on the nature of the dynamics of this vent

Deposits of this type, leading to formation of the layering at the Dais, are broadly similar to those described by Irvine (1987) in the Duke Island ultramafic bodies. Irvine (e.g., 1980) has experimentally studied density currents of this nature in some detail and delineated their role in cumulate processes, and his results may be applicable to the possible avalanche deposits that fed the layering at the Dais Intrusion.

### 9.4.3.5 Simulation of Sill Propagation with Lateral Flow

A prime feature of real mush column evolution clearly involves the lateral propagation of sills or sheets of magma. The leading tip and the chilled margins of large diabase sills are commonly phenocryst-free and are chemically the most differentiated and have the most spatially uniform compositions (e.g., Marsh, 1996). If a tongue of entrained phenocrysts is present, it is always found about a kilometer or more behind the sill tip. This suggests, as expected in the Simkin

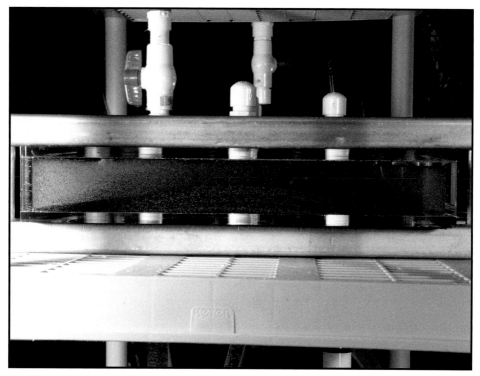

Figure 9.16 A similar view of layering but also showing the buildup of small suspended "dusty" particles at the ends of the tank, outside of the depositional area of the larger particles. When the system is allowed to go into repose, these fine particles settle to form a well sorted layer covering the larger particles.

Effect (Figure 8.11), that as the sill propagates ahead the entrained phenocrysts drop from the flow and advancement of the sill is by chemically differentiated, low viscosity (i.e., particle free), and highly mobile melt. The lateral margins are thus, in effect, cauterized by differentiated, crystal-free melt chilled from the advancing magma. Although the lateral ends of sills are notoriously elusive for observation, the extension of the Basement Sill eastward from the south lip of Bull Pass is exceptionally well exposed (see Figure 5.7), and the limit of the Opx Tongue material is also obvious here. The sill decreases from over 300 m thick to less than a meter over 7 km and, as mentioned in explaining this figure, the extension of the sill may have come about partly due to collapse of the pressure field driving the initial emplacement, squeezing the low viscosity, crystal-free magma ahead.

The important experimental question involves the rate and pattern of flow as the sheet extends laterally. That is, how is the rate of extension reflected in the distribution of particles behind the leading front of fluid? Are there telltale signs in the deposit of particles that tell whether the advance was steady or pulsate? Is this process reflected in the three-dimensional chemistry of the leading tip of the sheet?

This process is challenging to simulate experimentally, because the fluid in the lateral ends of the sill (i.e., tank) is largely stagnant unless there is a vent somewhere in the end of the tank allowing further fluid flow. This was simulated by passively withdrawing fluid through the end of the tank through a series of small vents attached to a fluid manifold connected to the main plumbing of the mush column. This allows the sill to advance laterally under hydrostatic pressure. And, because the flow rate is small, only the finer particles reached the end of the sill. The actual dynamics can possibly be unraveled by detailed sampling along the center of the Opx Tongue through to the crystal-free dolerite, where reactivation of melt in being squeezed ahead may contain clots of Opx Tongue material rather than individual, well-sorted Opx primocrysts. Clots of Opx material are found at various locations, especially near the Dais Intrusion (see Figure 7.7) and, besides indicating the strength of the flow, these may reflect this process of repeated impulses in the flow.

### 9.4.3.6 Time Series of Pressure Pulses

A primary feature of active volcanic systems is the strong variation in time of the eruptive flux (e.g., Wolfe et al., 1987). This must also be characteristic of the deeper magmatic system, and it may play a central role in the entrainment and transport of crystals through the erosion of deposits laid down under a weaker magmatic flux. The entire system may exhibit a certain resonant pattern of transport, deposition, and possibly eruption of crystals, which is established between the time series of applied driving pressure and the geometry of the input and exit vents at each level in the system. There are, in essence, two styles of pressure variations. One is a high frequency variation, characterized by variations on the scale of days, and the other is a low frequency or long-term variation that may characterize the eruptive output for an entire string of eruptive episodes over periods of 10s of years (e.g., Simkin, 1993).

These eruptive variations can be manifested experimentally through imposed short-term variations in the flow during a single episode, where there are always some particles in suspension and in motion. The depositional features so formed may be delicate, continuous, and pervasive. On the other hand, when the driving pressure is of the low frequency type, the full flow comes completely to rest, all particles are deposited, and then the flow restarts (the style of restart is as discussed in Sections 9.4.2 and 9.4.3). The final delicate layering of the previous cycle is strongly disrupted, and an erosion or entrainment surface is formed at key locations throughout the system. The size spectrum of particles in the erupted flow is apt to be decidedly distinct in each case. For high frequency variations in the ongoing flow, the size range of particles is expected to be much more restricted and smoother, whereas with the low frequency variations the size range is expected to be wider and more ragged.

It is this style of changes in the flow field that may have given rise to the striking changes in layering in the Dais Intrusion. That is, the power part is characterized by many smaller scales of delicate layering, primarily perhaps due to subtle changes in the incoming deliveries of slurry, leading to cascading and avalanching slurries causing pronounced sorting and layering (Bedard et al., 2007). The large "Event Horizons" on the other hand (see Figure 7.2, etc.), may be due to major hiatuses in the incoming slurry deliveries, indicating a low frequency series of driving pressure pulses.

### 9.4.3.7 Tracer Particles and Particle Tracking

When phenocrysts appear in lavas, it is commonly assumed that they have always been near each other. That is, it is assumed that neighboring crystals have nucleated and grown juxtaposed and have been transported similarly. Crystals have been lost from the flow, but on the scale of millimeters the surviving crystals have had similar histories, and thus a similar provenance. But this may not necessarily always be true. Detailed isotopic studies suggest that some magmas have experienced mixing and mingling on a very small scale such that even neighboring crystals may have experienced greatly different histories (Davidson et al., 2007). And U-Th disequilibrium suggests a series of processes where crystals and original melt have become disassociated by processes that are yet to be recognized.

This feature is also found in the Opx Tongue primocrysts that, as presented later, carry a markedly distinct isotopic signature over that of the crystal-free dolerite Carrier Magma (see Figure 10.6 and nearby discussion). In these experiments, individual identical (marked) particles also became separated after entrainment simply due to their placement within the flow and subsequent deposition and re-entrainment. Large scale isotopic mapping on the outcrop scale and from layer to layer may hence reveal how extensive this process may be, which may further indicate the sequence of deliveries and information on the nature of the ultimate source region.

### 9.4.3.8 Multiple Conduit Pathways

The conduits in real systems are apt to shift locations during prolonged magmatic activity or even to operate simultaneously at two or more locations at any level in the mush column, similar to what has been suggested by Muirhead et al. (2012, 2014) for the MCDV system. Smaller conduits become solidified, sealing them off, and new vents may form, or the flow becomes concentrated in re-activated, still mushy large conduits. This process presents the opportunity for the upward flow to optimize itself, presumably to minimize heat losses in the magma, and to transport and deposit particles in two or more competitive flow patterns. Even under a steady driving pressure, the flow may wax and wane locally. In these experiments, although solidification was not a factor, valves at each level were turned on and off periodically to stimulate and evaluate flow-competitiveness between various styles of circulation. Although both flows always stayed active, there was a marked tendency for the more direct or simpler flow to prevail, and the deposition of particles at each level interacted somewhat or interfered with one another, making a crude form of crossbedding. In nature, the weaker flow is likely to close off by solidification, leaving or enhancing the stronger conduit flow. Each of the previous areas of investigation mentioned is also affected by this property of the system.

### 9.4.4 Summary of Experiments

Although each series of experiments were intended or designed to reveal or investigate certain specific features of a working magmatic mush column, collectively they have much in common. The core feature is that of a particle laden fluid attempting to successfully traverse a certain style of magmatic plumbing, establish a series of sills, and perhaps break through to the surface, forming lavas and volcaniclastics. The concentration of the flow into a central conduit is essential to ward off the ever-ongoing effects of progressive solidification, which increases the local flux, allowing crystals to be entrained and transported throughout the system. This focusing of the flow field has a major thermal effect on the crust, leading to widespread heating, hydrothermal flows, and perhaps massive partial melting of the adjoining crust. This thermal insulation provides the opportunity for more extensive development of an intrusive or plutonic magmatic environment where substantial masses of entrained crystals can be deposited, leading to a wide assortment of crystal sorting and chemical differentiation.

In the end, perhaps the most important aspect of experiments of this type is that they display an abundance of processes and effects that, in turn, guide what to look for in the field, in the rock themselves, which is the final court of appeal.

# 10

## Opx Provenance

### 10.1 Introduction

The abundance and textures of the large grains of Opx forming the Tongue throughout the Basement Sill reflect the entrainment of these crystals essentially as sludge in the magmatic system. Among many major questions concerning the nature of these crystals, perhaps the most obvious concerns their provenance. That is, the ultimate location of their origin and textural development. Do these crystals represent foundered sub-crustal mantle wall rock or do they represent a deeper, integral part of the magmatic mush column that is rich in cumulates of this type. There are several perspectives useful in deciding this question, each of which indicates an uppermost mantle source with a continental crust heritage.

### 10.2 Rock Nature

The crystals themselves look old, not crystals that have been grown recently, and their size and abundance also make them unusually distinctive relative to all crystals nucleated and grown within the doleritic magma itself. The fine-grained dolerite texture is well known and distinctive, making these crystals highly anomalous. Although many solitary crystals are idiomorphic, others appear as clots where the internal common grain boundaries form ~120° junctions, indicating an earlier source rock sufficiently old to undergo strong textural annealing (Figures 7.7 and 10.1). And the sheer abundance of these crystals makes them more than simply an accidental or casually entrained component. The Basement Sill has a volume of about 1,100 km$^3$ and perhaps as much as one third of this can be ascribed to the Opx Tongue rock, which is nearly three times the volume of Skaergaard; the essentially monomineralic source material is vast in volume. The overall character of size, raggedness (at times), abundance, and position in the magmatic system, strongly resembles other systems dominated by xenocrystic–primocrystic mantle olivine as at Kilauea, Rum, Skye, and the mafic bodies in Southeastern Alaska.

195

Figure 10.1 A full thin section scan of sample A 270, an Opx cumulate from the Dais Intrusion, illustrating the apparent age and rounding of some of the grains, reflecting the slurry nature of the entrained collection of primocrysts

## 10.3 Phase Equilibria

Additional insight can be gained by examining these whole rock chemical data in the basalt normative phase system, as introduced in Figures 8.9 and 8.10, based on the works of Walker et al. (1979) and Stolper (1980). These data are plotted in the tetrahedral system Di-Ol-Sil-Plag and projected from the Plag apex (Figure 10.2). The full suite of analyses adheres closely to the experimental phase boundaries. Beginning at the Opx endpoint at the bottom boundary, the Opx Tongue rocks analyses trend directly to the Di-Opx-Plag cotectic and then closely follow the cotectic toward the Sil corner with the silicic segregations making up the field closest to the silica corner. That these rocks have exactly these phases points to the utility of this system in assessing phase equilibria in this suite of compositions. The accompanying inset diagram shows arrows on the phase boundaries indicating the direction of crystallization during cooling and also the movement of the phase fields with increasing pressure (1 and 1.5 GPa). This relatively tight pattern of analyses, assuming complete equilibrium, might suggest that the phase equilibria for all samples was set at relatively low pressures, perhaps no greater than 0.5 GPa (5 kb), which represents mid to lower crustal depths. As is clear from the petrography and will be subsequently further revealed via their isotopic character, however, these Opx primocrysts are in striking contrast with the governing basaltic "carrier" magma. These Opx crystals clearly did not crystalize from this basaltic magma, which is the governing magmatic fluid. The fact that these bulk rock compositions adhere to the phase boundaries in this system does not indicate chemical equilibrium, but rather that a mechanical mixture of these two principal components, mainly a tholeiitic basalt and Opx primocrysts, gives compositions inherently matching the defining compositions of this normative phase system. This is reassuring because by this basalt being bathed in massive amounts of Opx it

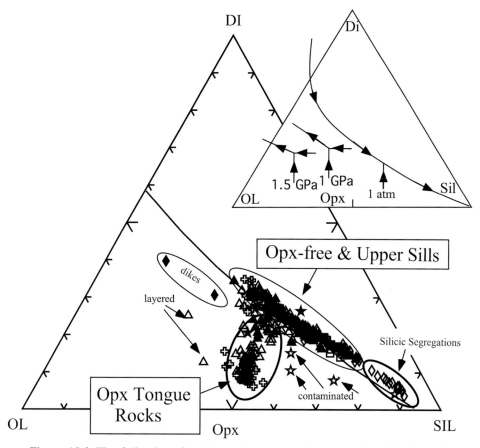

Figure 10.2 The full suite of whole rock analyses plotted in the CMAS basalt system of normative Di-Sil-Ol-Plag and projected from the plagioclase apex. (The symbols are the same as for Figure 8.1a.) There is a close overall adherence to the low-pressure phase boundaries, where the primocryst-free rocks closely follow the phase boundary toward the SIL corner where the silicic segregations are the final result, and the majority of the Opx-rich rocks diverge and follow the dog-leg phase boundary to the Opx composition. Two compositions from strongly layered compositions and some others from the Dais Intrusion fall significantly off this low-pressure boundary, which may reflect a higher-pressure equilibrium as indicated by the phase fields on the inset phase diagrams at the upper right. Overall, this reflects the strong roles of the Carrier Basaltic Magma being bathed in an abundance of Opx primocrysts.

has been forced to be saturated in Opx, which, as mentioned in Chapter 7, is also seen in the large abundance of fine grained Opx found in the Dais layered anorthosite units. Additional insight in this regard comes from the estimated temperatures of equilibrium of the Opx primocrysts.

We have previously published a vast array of crystal chemistry information on the Opx primocrysts (and all other phases) found in the Dais Intrusion (Bedard et al., 2007). Additional reconnaissance compositional data on Opx elsewhere in the Basement Sill are essentially identical to these Dais compositions. The typical temperatures estimated from these data, using the Ca in Opx and Fe-Mg exchange (Brey & Kohler, 1990), assuming a total pressure of 0.1GPa (1 kb), are consistently near 1,200–1,250°C. This total pressure is clearly a minimum value as the Dais itself was emplaced at a depth of 4–5 km in the crust (~0.15 GPa), and, unlike the tiny Opx crystals, which certainly grew at this depth, the large Opx primocrysts have a much deeper provenance. Although the pressure effect on these geothermometers is not large, increasing the assumed pressure to 1 GPa (10 kb) increases the temperature estimates by about 50°C. The bulk magma overall probably did not generally exist at these temperatures, but that the magma was of high temperature may be indirectly inferred by the extensive melting of the granitic wall rock found in the Bull Pass region, where the central feeding zone is located. Judging from these temperatures alone, with a geothermal gradient even as high as 40°C per km suggests a depth of over 30 km.

It is difficult to directly estimate a pressure of equilibrium for the Opx primocrysts because the full phase assemblage in the source region is not known. That is, the massive sludge of orthopyroxenite is mainly just Opx with a subordinate amount of Cpx and a small concentration of plagioclase. The amount of Al in the Opx is generally 1.2–1.8 wt.%, similar to Opx in Stillwater and Dufek, which is not inordinately high, but if this was the CMAS (Ca-Mg-Al-Si) system containing garnet or spinel and with similar amounts of chrome, these levels of Al might be at pressures as high as 1.5 GPa or more (e.g., Klemme & O'Neil, 2000). In somewhat similar, but pegmatitic, leucogabbronorites in Finland, the Al contents are much higher (up to 6–7 wt.%) for estimated pressures of ~1.5 GPa. The plagioclase in this assemblage is much less anorthite rich (~$An_{50}$), whereas here it is consistently near $An_{85}$, reflecting the low overall sodium content, which competes for Al and thus affects the Opx Al content (Heinonen et al., 2020).

Although the Dais Intrusion does not contain Fe–Ti oxides, the Dufek does and they give a corresponding temperature of 1,100°C and an oxygen fugacity near that defined by the quartz–fayalite–magnetite buffer reaction (Himmelberg & Ford, 1977). At this oxygen level, these oxides appear late in the crystallization sequence and the prevailing emplacement temperature might well have been similar to that of the Dais (i.e., ~1,200°C).

The occurrence of mainly mono-mineralic primocrysts is a fairly common situation in large magmatic systems. At Kilauea, for example, the picritic lavas are laced with fragments of large forsterite-rich olivine (~$Fo_{85}$, Murata & Richter, 1966; Garcia et al., 2003; Marsh, 2013) that is in strong disequilibrium with the carrier basaltic magma. Primocrystic olivine of this composition is pervasive in many picritic systems and is also commonly found in kimberlites, where olivine closer to $Fo_{92}$ might be expected (Arndt et al., 2010). At Kilauea, there is also no sign of other phases that might indicate the full nature of the peridotitic source rock; the coexisting cpx, opx, and plagioclase or spinel have evidently all dissolved into the basalt. Remarkably enough, however, on an oxide fractionation diagram, these olivine primocrysts appear as perfect fractionates, and on CaO versus MgO diagrams they give tight olivine-control variations (Wright, 1971; Marsh, 2013), just as the Opx primocrysts do here (see Figures 8.4 and 8.5).

## 10.4 Isotopes

A number of previous studies have revealed the notoriously diverse Nd–Sr isotopic character of the Ferrar Large Igneous Province as a whole (Hergt et al., 1989a; Fleming et al., 1995; Elliot et al., 1999; Elliot & Fleming, 2004, 2017, 2021). For example, initial $^{87}Sr/^{86}Sr$ ranges from ~0.71050 to 0.71270, whereas at mature spreading centers MORB is much more uniform over long distances, at ~0.7025. After extensive hydrothermal interaction far from the ridge this commonly increases to ~0.7037–0.7040, which defines a distribution much tighter than that commonly found in a single locality in the Ferrar Large Igneous Province (LIP). It is thus of prime importance to gauge whether this isotopic diversity or heterogeneity exists mainly on a regional scale or on a local scale within individual otherwise apparently homogeneous sills, like the Peneplain Sill, and within the Opx-dominated rocks of the Basement Sill, as at the Dais Intrusion. The Peneplain Sill profile analyzed here is from near Pandora's Spire at Solitary Rocks, which is the antithesis of the Dais Intrusion, containing no phenocrysts upon emplacement, is texturally uniform from top to bottom, and looks in all respects homogeneous throughout. For both this Peneplain Sill locale and the Dais Intrusion, initial $^{87}Sr/^{86}Sr$ and $^{143}Nd/^{144}Nd$ were measured and reported by Foland and Marsh (2005). To further augment this analysis, oxygen isotopic determinations were also made on whole rock and mineral separates (Opx and plagioclase) in the Dais samples (Larson & Marsh, 2005).

### 10.4.1 Strontium

A histogram (Figure 10.3) of all the measured values of initial $^{87}Sr/^{86}Sr$ reveals a highly radiogenic suite, variable in initial calculated $^{87}Sr/^{86}Sr$ (~0.7106–0.7122). The average is almost identical to the averages (0.7112) for the dolerites of South Victoria Land and Tasmania and is only slightly more radiogenic than the Dufek rocks as compiled by Faure and Mensing (2010, pp. 419, 451). The spread in values for each unit is significantly larger than that (~0.7090–0.711) of the Portal Peak sill found by Hergt et al. (1989a). As argued by others

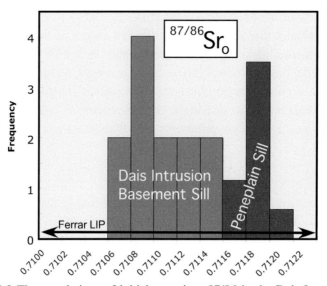

Figure 10.3 The populations of initial strontium 87/86 in the Dais Intrusion and Basement and Peneplain Sills (for an age of 183 Ma). Although the magmas are coeval, contiguous, and spatially near one another, the strontium populations are distinct and variable, and the spread in abundances nearly cover that as found for the entire Ferrar Large Igneous Province, as indicted.

(e.g., Fleming et al., 1995) and as will become apparent here, these strong variations are not due to alteration. Each body shows a wide range of $^{87}Sr/^{86}Sr$, collectively covering over 50 percent of the entire range of Ferrar LIP (as indicated) as so far known. Moreover, even though these units give every indication, geologically, of being contiguous, their initial $^{87}Sr/^{86}Sr$ are each distinct from one another. When the values for the Peneplain Sill are plotted stratigraphically, the pattern suggests that this sill was formed by near simultaneous injections of three isotopically distinct batches of magma (see Figure 10.4). Although this is reminiscent of the typical pattern of successive emplacements found in other sills, like the Insizwa Sill (Ferré et al., 2002), Beacon Sill (Zieg & Marsh, 2012), and the Basement Sill (Charrier, 2010), it is surprising that the successive batches of magma are isotopically distinct. This implies that the emplacement sequence did not simply start, stop, and start again, but that the new, chemically distinct magma was a fresh batch from deeper or elsewhere in the system. And since there is only a weak variation in MgO in the sill itself, there is also a similar weak correlation between initial $^{87}Sr/^{86}Sr$ and bulk MgO (not shown). A fairly convincing inverse correlation between initial $^{87}Sr/^{86}Sr$ and bulk MgO for the whole FLIP has been suggested by Elliott and Fleming (2017), where the most radiogenic lavas and sills have MgO of ~3 wt.%, which can be more readily evaluated by the large variations in MgO in the Dais Intrusion.

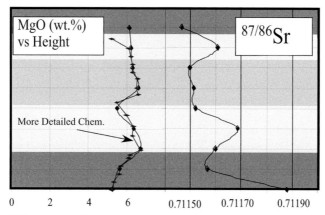

Figure 10.4 The composition (MgO) and initial Sr-87/86 in the Peneplain Sill in a sampling profile taken at Pandora's Spire in Solitary Rocks. The sampling profile was repeated to give it more detail in MgO. Although the sill itself appears remarkably homogeneous in the field, the variation in initial strontium is highly significant. The variations in Sr-87/86 and MgO also suggest that the sill was emplaced in several episodes, which are indicated by the shading.

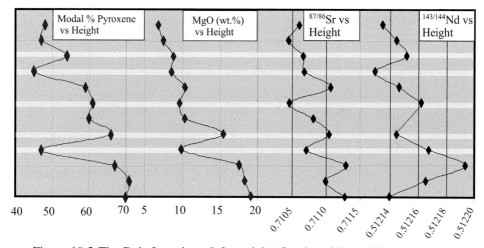

Figure 10.5 The Dais Intrusion: (left to right) Stratigraphic variations in modal mineralogy, MgO content, initial Sr-87/86, and neodymium 143/144. The variation in MgO is closely correlated with modal Opx, as is Sr-87/86, whereas Nd-143/144 has a mixed correlation.

$^{87}Sr/^{86}Sr$ variations in the Dais profile show a remarkably detailed positive correlation with MgO and modal pyroxene (Figures. 10.5 and 10.6). The basal ultra-mafic, Opx-rich units with ~20 wt.% MgO are by far the most radiogenic (~0.71160) and the upper tholeiitic units with ~7 wt.% MgO are the least radiogenic (0.71050). The ratio of modal Opx to plagioclase is clearly a major

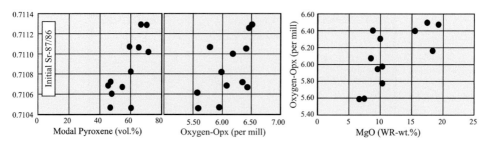

Figure 10.6 The Dais Intrusion: (left to right) Positive correlations between Sr-87/86 and modal pyroxene, Sr-87/86 and Opx $\delta^{18}$O, and Opx $\delta^{18}$O and whole rock MgO content. The Opx rich rocks are more radiogenic in Sr-87/86 and are also richer in $\delta^{18}$O.

factor in controlling these variations; the tiny plagioclase grains grew late in the ascending magma and thus mainly characterize the isotopic nature of the Basaltic Carrier Magma. The large entrained Opx (and ancillary Cpx) cumulate crystals, on the other hand, characterize a distinct $^{87}Sr/^{86}Sr$ source environment. In this regard, isotopic measurements by ion probe (SIMS) on individual small plagioclase crystals show characteristic variations depending on their proximity to the large Opx grains (Mathez et al., 2005). Grains nearly touching Opx crystals have the isotopic signature of Opx, and nearby grains 2 or 3 mm away have the isotopic signature of the Carrier Magma. This strongly suggests an inter-diffusional relationship between the Opx and the Carrier Magma. The values of $^{143}Nd/^{144}Nd$ are less distinctive, showing a more limited range (~0.51216–0.51212, mostly), and it is also partly correlated with Sr and bulk rock composition.

### *10.4.2 Oxygen*

To characterize further the chemical identify of the source materials, oxygen isotopic ($\delta^{18}$O) determinations were made on these same Dais samples for whole rocks and mineral separates (Figure 10.6; Larson & Marsh, 2005). Opx primocrysts show 6 ± 0.5 per mil variations (0.1.per mil uncertainty relative to SMOW) and plagioclase gives slightly higher values (6.25 ± 0.5). As in the earlier correlations between $^{87}Sr/^{86}Sr$ and MgO, there is a fairly close correlation between whole rock $\delta^{18}$O and $^{87}Sr/^{86}Sr$ in the lower, ultra-mafic part of the Dais and much less correlation in the upper part, which is dominated by the Basaltic Carrier Magma. There is no clear correlation between oxygen values for plagioclase and Opx; the indistinct overall correlation between oxygen and initial $^{87}Sr/^{86}Sr$ eliminates the concern of serious late-stage hydrothermal effects. There is a moderate positive correlation between Opx oxygen and whole rock initial $^{143}Nd/^{144}Nd$, but none between plagioclase and $^{143}Nd/^{144}Nd$ (not shown). Oxygen fractionations between plagioclase and Opx are in the range −1.38 to 0.87, which is unusual. Whole rock oxygen is also elevated beyond the usual 5.7–5.8 per mil values of MORB. Overall, the strongly radiogenic $^{87}$Sr/$^{86}$Sr of these rocks coupled with elevated oxygen suggests magma either variably and extensively contaminated with continental crust, which has often been suggested for Ferrar Dolerites, or that this magmatic mush has been formed from a more deliberate mixture of older sub-continental orthopyroxenite and rift-related dolerite magma. The late plagioclase clearly reflects growth from magma isotopically distinct from that giving rise to the massive Opx phenocrysts.

As might be expected, the oxygen in Ferrar doleritic rocks is also commonly highly variable; a comprehensive review is given by Elliot and Fleming (2021). In the two sills at Roadend Nunatak, $\delta^{18}$O varies between 6 and 7‰ in the lower sill and between 5 and 7‰ in the upper sill, with no systematic or characteristic pattern, as is also found for the lowermost Mt. Achernar Sill (Faure & Mensing, 2010, pp. 431, 437). Although most $\delta^{18}$O measurements are for whole rocks, values for cogenetic plagioclase and pyroxene from the middle Mt. Achernar Sill show a close correspondence from about 1.5 to 5.5‰ at the lower contact to 4 to 6‰ in the interior (Faure & Mensing, 2010), suggesting temperatures of equilibration of ~600°C at the base to ~1,000°C in the interior, which are entirely reasonable; the theoretical contact temperature is ~625°C. And measurements of plagioclase separates from a single lava flow with large $\delta^{18}$O variations gave a tight mantle-like signature (Fleming et al., 1992).

Whole rock measurements of $\delta^{18}$O of the Dufek Intrusion cumulates of gabbroic, leucogabbroic, anorthositic, and pyroxenitic composition are between 6 and 7‰, whereas rocks from the Forrestal section containing the thick upper granophyre and a spectrum of other lithologies including felsic dikes, gabbros, and diorites from near the contact near Mt. Lechner, and anorthositic inclusions in gabbro give values from 0 to 6.1‰ (Ford et al., 1986; Kistler et al., 2000). Mineral separates generally reflect these same distributions and $\delta^{18}$O fractionations between pyroxene and magnetite suggest magmatic temperatures of 1,320 to 800°C. If there is any generality to be gained from these Ferrar $\delta^{18}$O results it is that lower values, especially for the Dufek-Forrestal rocks, reflect high temperature alteration, most likely from an in situ hydrothermal system induced by the thermal regime of the intrusion itself. This is unlike Ferrar lavas that show higher $\delta^{18}$O, reflecting low temperature alteration. The lack of any clear correlation between oxygen and initial Sr suggests that contamination is a separate effect (e.g., Hergt et al., 1989a).

## 10.5  Isotopic Indications

The common problem in ascribing a cause of the significant variations in $\delta^{18}$O and initial $^{87}$Sr/$^{86}$Sr is partly due to inadequate petrologic controls on what is

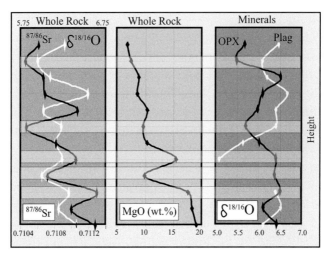

Figure 10.7 Dais Intrusion: (left to right) Stratigraphic variations in whole rock initial Sr-87/86 and $\delta^{18}O$ (plotted together), whole rock MgO (wt.%) content, and last $\delta^{18}O$ in mineral separates of Opx and plagioclase. There is a fairly close positive correlation between whole rock Sr-87/86 and $\delta^{18}O$, along with the overall positive correlation with MgO, which lessens with height. $\delta^{18}O$ for Opx and plagioclase show a slight positive correlation but also indicate that the Opx is less well correlated with MgO, which may indicate some diffusional contamination with the host magma (see text).

controlling these variations in specific intrusives, which, due to burial and crystal size, have escaped low temperature alteration effects. In this respect, given the strong variation in rock type, the Dais rocks are well suited to identifying the ultimate cause, contamination versus alteration, responsible for these effects. Plots of $\delta^{18}O$ versus MgO content (proxy for pyroxene), initial $^{87}Sr/^{86}Sr$ versus modal pyroxene, and initial $^{87}Sr/^{86}Sr$ versus $\delta^{18}O$ (Figure 10.7) each show fairly good positive correlations. Just as in controlling the whole rock composition and fractionation, it is the large Opx sludge material that controls the fundamental variations in these isotopes. And, unlike for Ferrar lavas (Elliot & Fleming, 2017), here there is a positive correlation between initial $^{87}Sr/^{86}Sr$ and MgO. These correlations might be made even tighter if the Opx crystals were hand-picked by size to include only the largest crystals. That is, it is clear that the Carrier Magma is less radiogenic in Sr and has near MORB-like $\delta^{18}O$ (~5.60 per mil) such that, in being bathed in this crystal-rich melt, smaller Opx crystals might already be "diffusion-altered" by the magma, and vice versa. That the $\delta^{18}O$ is overall slightly heavy, over MORB values, and the initial $^{87}Sr/^{86}Sr$ enriched, might also suggest that this is the actual nature of the Opx source material giving rise to these Ferrar-McMurdo Dry Valleys magmas, which has been inherited much earlier. In the reverse of the Dais pyroxene, the pyroxene of the Dufek is less radiogenic (Kistler et al., 2000), which very likely indicates that these are newly grown crystals. (The obvious idea that a small amount of subducted pelagic-like sediment might serve the purpose of increasing each of these isotopes prior to pyroxene formation in the Dais source rock has other difficulties.)

In terms of building a new continent or producing continent-like materials, Iceland is a prime present-day example, the whole landmass being less than about 20 Ma old. Almost all the lavas generated within present day volcanic belts in Iceland are anomalous in $\delta^{18}O$, ranging from 0.0 to 6.1‰, being both systematically lower and more variable than unaltered volcanic rocks elsewhere (e.g., Hattori & Muehlenbachs, 1982; Gunnarsson et al., 1998). As in the Ferrar, this depletion is commonly attributed to either assimilation or partial melting of crustal rock that has been hydrothermally altered at relatively high temperatures >250°C by low O meteoric ground waters (e.g., Taylor, 1977, 1987; Hattori & Muehlenbachs, 1982; Oskarsson et al., 1982). Even for fresh, recent silicic lavas, the variation in $\delta^{18}O$ in a single volcanic center varies from 3.5 to 4.5‰, all being lower than expected MORB values (Gunnarson et al., 1998). And $\delta^{18}O$ values over a relatively small geographic area can vary beyond this range. This variable signature of $\delta^{18}O$ in lavas within the same geographic area suggests strong variations in the nature of the hydrothermal systems giving rise to these variations, either in size of the system, strength of the system, or the longevity of the circulation. This is certainly reflected in the wide range of $\delta^{18}O$ values found at the Dufek-Forrestal Intrusion. On the other hand, in these same Icelandic lavas the range of values of $^{87}Sr/^{86}Sr$ and $^{143}Nd/^{144}Nd$ are remarkably small, 0.70325–0.70337 and 0.51297–0.51300, respectively. The relatively small range of values of $\delta^{18}O$ for the Dais rocks, which are certainly voluminous, points to a source region that is apparently relatively homogeneous in $\delta^{18}O$.

### *10.5.1 Osmium*

In hopes of identifying the ultimate source of the Ferrar LIP magmatism in terms of a plume or simply local mantle rift-related melting, the concentrations of PGE and associated Os isotopes ($^{187}Os/^{188}Os$) have been measured for the Dias and a section of the Basement Sill in West Bull Pass (Choi et al., 2019). The $^{187}Os/^{188}Os$ values range from 0.1609 ± 0.003 (2σ) to 8.100 ± 1.600 (2σ), confirming a previously published estimated initial $^{187}Os/^{188}Os$ value for Ferrar magmas of 0.145 ± 0.049 (2σ) (Brauns et al., 2000). These values are unlike those of depleted asthenosphere and lithospheric mantle. The patterns of PGE abundances exhibit positive, convex-shaped slopes between the IPGE (Os, Ir, and Ru) and PPGE (Pt, Pd, and Rh), unlike for the roughly chondritic PGE abundance patterns of lithospheric mantle, but more similar to arc-type mantle material. Unlike for O and Sr, there is no strong correlation of PGE concentrations and whole rock MgO content. But perhaps the most significant feature of these measurements is the extreme sub-chondritic Os/Ir ratios (<0.33), regardless of rock type, much unlike typical plume-derived magmas where Os/Ir = ~2.0. Choi et al. (2019) suggest instead that these low Os/Ir values are more consistent with simple rift-style decompression melting along a fossil subduction zone where the mantle involved had been previously flux-hydrated by lithosphere dehydration during prolonged subduction associated with super-continent assembly. Regional mantle hydration might thus allow for rapid voluminous melting upon the pressure release of rifting and strong adiabatic ascent of the hydrated mantle.

One difficulty with this scenario is that the Ferrar sills and dikes show no unusual role of volatiles in their emplacement and evolution. Miarolitic cavities, trapped and/or filled vesicles, and exhalative effects in the roof rocks in the form of pebble dike formation and other signs of phreatic activity are truly more than scarce. The phreato-magmatic activity at the highest levels, such as at Allan Hills, primarily indicates magma interacting with ground water. And significant quantities of water also promotes upon ascension and saturation heavy bursts of nucleation as the volatile depressed phase boundaries are unavoidably encountered (e.g., Marsh & Coleman, 2009), which is not seen in even the highest level sills, for example, at Mt. Fleming. The dolerite magmas seem dry, perhaps containing ~0.25 wt.% water, similar to that of MORB and Hawaiian magmas.

## 10.6 Conclusions

From the basic nature of the style, abundance, and geochemical characteristics of the Opx primocrysts of the Basement and Peneplain Sills there is every indication that they have been entrained from a subcontinental source that is massive and widespread. As a class in and of themselves they are somewhat remarkable in their apparent uniformity, however anomalous relative to typical olivine-dominated mantle material, and as will be further shown in Chapter 12, apparently represent a truly extensive source rock in the sub-Antarctic mantle of 183 Ma. In a study of the anomalous oxygen isotope composition of Karoo and Etendeka picrites, Harris et al. (2015) conclude that the mantle source region itself is anomalous in its isotopic oxygen, that it has been like this possibly since Bushveld times, and that it may reflect a strong eclogite component from emplacement of oceanic lithosphere into the cratonic keel during Archaean subduction.

The Basaltic Carrier Magma, on the other hand, is remarkably anomalous in its siliceous character and widespread isotopic heterogeneity, the origin of which will be further considered in Chapters 13 and 14.

# 11

# Noritic Magma, Primocryst Entrainment, and Source Sampling

## 11.1 Introduction

To further understand the possible ultimate source of the massive slug of Opx in the Basement Sill, it is essential to appreciate something of how large masses of foreign crystalline subcontinental material might become intimately entrained in magma. Other than the obvious importance of the material itself, it also reveals how large samples of the underlying crust and mantle are magmatically excavated. The extensive outpouring of this material is clearly not simply a unique accidental occurrence. The key resides in the deeper style of magma transport, the fixedness of the plumbing system, and the longevity of the overall system. The style of sill emplacement in the Dry Valleys reflects the regional stress field, the consistency of the crust, and the power of the magmatic event.

## 11.2 Rifting and Sill Emplacement

In the global Gondwana rifting event, many similar sill systems came about and each one is somewhat distinctive. Just as along ocean ridges, each ridge segment has basins or centers where the magmatism is centered. In the Dry Valleys with the formation of the incredibly uniform and sharp regionally expansive Kukri Erosion Surface or Peneplain, due to crustal erosional unloading delicate extensive exfoliation or expansion joints formed everywhere parallel to the Peneplain. The Basement Sill relentlessly follows these features everywhere except in and around the Bull Pass eruptive center. The thick stack of Beacon Supergroup sediments above the Kukri Peneplain were, for the most part, well consolidated and the bedding planes provided strong stress guides for the early sills to follow (Barrett, 1991). At the higher levels, where the sediments and earlier volcanics were less consolidated, the influence of the stress field became weak or neutral allowing haphazard magma emplacement (e.g., Muirhead et al., 2012, 2014). Thin sills forming just beneath the surface here migrate upward in places to form dikes cutting the overlying volcanics. As mentioned at the outset in Chapters 2–4, some small Ferrar sills and dike-like bodies can be traced almost continuously upward throughout the transition from dolerite to tephra (Grapes et al., 1974; Korsch, 1984; Muirhead et al., 2012). As mentioned, this tephra forms an extensive subaerial phreatomagmatic deposit known as the Mawson Formation, which is up to 170 m thick (e.g., Gunn & Warren, 1962; Gunn, 1966; Elliot & Larsen, 1993). The capping Kirkpatrick flood basalts are the volcanic equivalent of the uppermost dolerite sills (e.g., Fleming et al., 1995).

There is good reason to believe that this basic intrusive stack-like structure of sills may continue downward to much deeper levels in the crust. A thick stack of dolerite sills of a similar overall geometry, for example, is found in the high-grade metamorphic terrain of the southwest Scottish Highlands, which originally existed in the middle and lower crust (Skelton et al., 1995). And vertically extensive sequences of sills are imaged in deep seismic reflection studies in the North Sea, extending to and beneath the Moho (Hansen & Cartwright, 2006). These sills also emanate from central areas but are much more saucer-shaped, reflecting the stress field and state of consolidation and structural attitude of the governing sedimentary column. Broadly similar saucer-like features are also found deep in the continental margin in the North Rockall Trough northwest of Scotland (Thomson & Hutton, 2004), in the Karoo system of South Africa (Malthe-Sørenssen et al., 2004; Polteau et al., 2008a, 2008b; Neumann et al., 2011), and also in Britain (Goulty, 2005), which, like the Ferrar, is remarkably free of major dike systems leading to sill formation. Instead, the latter authors present an in-depth study of the time history of sill emplacement interpreted using extensive AMS (Anisotropic Magnetic Susceptibility; Ferré, 2009) measurements, offering the following as the chief mode of sill formation consisting of repeated cycles of: (1) injection of magma; (2) formation of a saucer-shaped sill; (3) pressure build up; (4) fracturation and pressure drop; (5) channeling of magma; (6) injection of (new batch of) magma; and (7) formation of a new saucer-shaped sill injection of magma until the magma supply stops. The effects of solidification, although not specifically mentioned here, are also a major factor in enhancing this emplacement sequence (Currier & Marsh, 2015). Solidification, in effect, throttles the emplacement process, producing a diverse emplacement center with anastomosing petals or sheets of magma reaching out from a central locus, much as is seen at Bull Pass.

This is also found in the saucer sill complexes, where the saucer shape reflects the heterogeneous stress field generated in the local crust due to sill emplacement. They also suggest, depending on the consistency of the overburden, this will be the general form of sills that are longer than two or three times their depth. If so, the great extent (>~100 km) and flatness of the Basement Sill likely reflects the rigidity of the crust due not only to the strong consolidation of the Beacon sediments, but also due to the additional stiffness added to the crust from the earlier emplacement and relatively rapid solidification (~1,000 years each) of the upper sills. Adding these upper sills to the crust is like adding a series of load-bearing structural plates; further suggesting that the buildup of sills was from the top down. This distinctive characteristic of Ferrar sill flatness in the Dry Valleys region may not be necessarily true in the Beardmore region (Grindley, 1963; Faure & Mensing, 2010, see their figure 13.2).

Given these observations, a stack of sills can certainly develop at nearly any depth as part of a magmatic mush column (Marsh, 2004). The sheet of wall/roof rock between successive sills thus becomes, in effect, an unsupported cantilever beam or sheet, which is unstable and upon collapsing dumps large amounts of country rock into the sill magma. In some ways this is akin to stoping, but the key difference is that at sufficiently high temperatures the stoped material disintegrates and becomes a source of primocrysts. That is, at sufficiently high temperatures the sill magma solidifies slowly if at all, allowing ample time for the cantilever to collapse, and the magma does not quench around the ingested materials, unlike in the crust where stoped blocks are coated in quenched magma and become caught up in the mushy solidifying magma (Marsh & Coleman, 2009). At Finger Mountain in the upper Taylor Valley, for example, sinking stoped blocks are suspended in the sill magma (Figure 11.1).

## 11.3 Primocryst and Bulk Source Ingestion

The actual physics of primocryst ingestion, reworking, sorting, entrainment, and re-deposition has been analyzed in some detail by Bergantz et al. (2017), where movies of the ongoing processes are exacting, explicit, and highly informative. A general characteristic of this overall process is that, with the establishment of an ascent or mush column, magma repeatedly traverses and processes the primocrystic material to a degree that only the most refractory minerals survive, eventually to become carried upward in the later magmas. That is, early magma traversing this region does so without significant interaction with the wall rock as the hotter magma quenches, cauterizing against the walls. Only in the later stages, after establishing a well-used, thermally enhanced mush column, do sill stacks develop and the entrainment process begins. Prolonged heating breaks down the earlier thermally cauterized contacts and wall rock sloughs off into the magma. This is reflected in the Dry Valleys dolerites by only the very last magma, mainly the Basement Sill containing entrained mineralogic debris, existing as single crystals and clots of crystals rather than as large chunks of country rock. In sum, then, this is a process most likely to have operated in the upper mantle below the

Figure 11.1 Stoping of blocks of Beacon sandstone at Finger Mountain. Large blocks near the roof are trapped in upper solidification front, some much smaller blocks appear deeper in the sill, which itself has an internal structure strongly suggesting multiple injections and earlier incorporation of the smaller blocks.

crust at high temperatures and at primarily fixed locations, similar to that operating in western North America (Bachmann & Bergantz, 2008) and beneath Hawaii today.

The Dry Valleys and Dufek magmatism each collected huge samples of the underlying source regions of, respectively, ultramafic pyroxenite and gabbro. The earlier Dufek magmas are mainly massive classic gabbros, broadly similar to those found at Stillwater and Bushveld, along with subordinate Opx-rich slurries and distinctive anorthositic deposits. The largeness of the system, reflecting its power and longevity, also had an unavoidably pronounced thermal effect on the local crust with the production of a major volume of rhyolitic magma as granophyre, which is also found at similar outpourings such as Bushveld. This silicic magma is clearly not a differentiate but partial melts from the underlying crust (e.g., Marsh, 1989b). The Dry Valleys system did not have as pronounced an effect on the granitic crust but is instead dominated by massive sampling of an Opx-rich deep country rock. This latter sampling was particularly clean in that there is little sign of anything but this major expanse of pyroxene-rich rock. As discussed in Chapter 6, however, this magmatism might well have produced deeper, voluminous crustal melting, as evidenced by the extensive high-level melting of the Orestes Granite in Bull Pass, but it was evidently not extensive enough to become mobilized into ignimbrites and rhyolites (see Currier & Flood, 2019).

## 11.4 Norite Production

These highly localized magmatic outpourings, Dry Valleys and Dufek, have delivered voluminous samples of an ultramafic to gabbroic source region that is distinct from the usual deep peridotitic mantle that monotonously delivers tholeiitic basalt like MORB. Ocean ridge tholeiites show close phase equilibrium adherence to olivine control (e.g., Walker et al., 1979), as do the Ferrar dolerites to Opx control. Stillwater and Bushveld are similar in this respect to the Ferrar, along with many other doleritic systems, in exhibiting little to no sign of having been spawned by a typical peridotitic mantle.

Yet, that similar volumes of noritic magmas of broadly this type can be produced wholly in older continental crust is clear from the Sudbury impact structure. Here, in a matter of a few minutes 1.85 Ga ago, a ~12 km bolide opened a transient crater 90 km wide and 30 km deep, that rebounded into a 200–250 km crater 3 km deep filled with ~30,000 km$^3$ of superheated melt (~1,700°C), all produced by wholesale melting of the crust (e.g., Zieg & Marsh, 2005, and references therein). The entire initial event took perhaps 10–15 minutes, with subsequent full solidification taking ~$10^5$ years. The resulting magma was a viscous emulsion, a mixture of molten dollops of granitic crust of all sizes bathed in a noritic magma. Much as a salad dressing emulsion, the entire mass soon (~10s of years) cleanly separated according to density into two nearly homogeneous layers: a granophyric (~70% $SiO_2$) upper layer (~1.75 km) and a noritic (~56% $SiO_2$) lower layer (~1 km), separated by a Transition Zone of mixed parentage where lithologic stuff collected due to neutral buoyancy from above and below. The two principal layers show an isotopic identity indicating a once intimate detailed interconnection, which came about by the intensely vigorous thermal convection in the initial superheated magma.

Now, the original Middle Proterozoic crust was certainly highly heterogeneous, containing regional doleritic dike swarms and sills along with a large variety of granitic lithologies and perhaps even some ultramafic bodies. The initial bulk compositional and isotopic repository, judging by the remaining original country rocks, was vast and varied, but the enormous energy associated with the impact ignited a series of magmatic process that both fully reorganized the crust and homogenized the isotopic records in the magmatic units produced. The crust was excavated right to the Moho, but there is no clear sign that the mantle was directly involved. This noritic magma produced by complete remelting of the lower continental crust is in some respects similar to that typical of the Ferrar system, but in detail it is far different. Nevertheless, because of the large volume of the sampling it is a valuable record of the type of noritic magma that can be produced solely within continental crust.

A chemical comparison of the Ferrar with Sudbury exemplifies this contrast (Figure 11.2) in the form of classical AFM and CMAS plots. The prevailing mafic

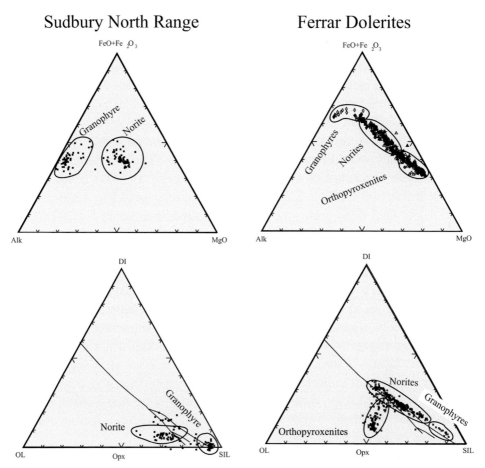

Figure 11.2 Comparison of the bulk rock compositions of the Sudbury Impact Structure, representing the wholesale melting of the crust, and the Ferrar dolerites of the McMurdo Dry Valleys. The upper pair of diagrams are AFM plots, and the lower pair are CMAS normative basalt phase equilibrium diagrams. In each case the Ferrar dolerites show a tight cohesive pattern reflecting a common parentage and coeval evolution, whereas the Sudbury rocks indicate a much more haphazard magmatic origin and evolution.

magma in each case is noritic, but the Ferrar AFM compositions form a tight, well-defined continuous trend beginning with the orthopyroxenites of the Basement Sill, continuing to the true norites defined by the upper sills (Peneplain, Asgard, Mt. Fleming) and ending with the granophyres from the lenses within the sills. The Sudbury AFM compositions are much more scattered, showing up as two distinct groups representing the massive upper granophyre and the basal norites, with no obvious genetic connection. The CMAS plots show similar distinctions: the Ferrar, as discussed in Chapter 8, closely adheres to the low-pressure phase boundaries, even to the extent of the ultramafic orthopyroxenites defining the well-known Opx dogleg. The Sudbury compositions show no adherence to any of the phase boundaries but are scattered; there is no sign that these compositions might be related through any process of protracted fractional crystallization. The Sudbury samples represent an excellent example of the bulk composition of perhaps typical continental crust, but it is unsuitable as a source material for any of the Ferrar materials. Moreover, even though the Ferrar basic magmas are regionally variable in bulk, trace element, and isotopic composition, from the Kirkpatrick lavas through to the sills, they are compositionally much more cohesive (e.g., Elliot & Fleming, 2017, 2021) than what can be produced wholly within continental crust.

# 12

# Regional Distribution of Ferrar Magmatic Centers

## 12.1 Introduction

The picture that emerges from the spatial distribution of overall sill bulk composition and Opx primocryst distribution in the Dry Valleys is that of a major, spatially restricted magmatic center, delivering as intrusives and subordinate lavas as much as perhaps 10,000–15,000 km$^3$ of magma over a remarkably short period of time. Judging from the inter-sill contacts and cooling times, taking perhaps only 10,000–50,000 years. The spatial distribution of sill lobes, striking primarily North and South, is broadly similar to what is found along ocean ridges where magma delivery is concentrated at specific locations and is then transported laterally along the ridge axes; a thorough review is given by Batiza (1996). Ocean ridge segmentation can occur at many scales from that defined by Transform Faults, on the order of 100s of kilometers, to that defined by the melt lenses along the central axes, which may be on the scale of ~10 km (e.g., Sinton & Detrick, 1992). Any possible segmentation in the Ferrar system will be reflected in the spacing of emplacement centers along the Transantarctic Range. This may be recognizable as evidently this nascent rift system, defined today by the Ferrar Large Igneous Province, became stranded after initiation, and the rift continued developing elsewhere, similar to that of the Eastern North America (Philpotts, 1992) or the much older Mid-America Rift System (e.g., Hinze & Chandler, 2020).

## 12.2 Greater Ferrar Magmatic System

During the span of time of formation of the full Ferrar system, taking perhaps ~1 Ma or less, given a full spreading rate of 10 cm/year, the surviving rift would have reached a width of 100 km, certainly becoming fully oceanic, dominated by MORB type basalts generated from typical peridotitic mantle source rock with olivine-controlled fractionation. So, this preponderance of Opx dominated primocrysts is a key indicator of magmatic–tectonic transition from continental to oceanic influences and also of the overall style of deeper magmatism during rift initiation. Assuming other similar magmatic centers developed along the Ferrar rift, the subcontinental mantle can be sourced from outpourings similar to the Ferrar Dry Valleys system.

### 12.2.1 Dufek Intrusion

The only other obvious center of this size and scope is the massive Dufek Intrusion (~6,600 km$^3$; Ferris et al., 1998, 2003), some 2,500 km away in the northern part of the Pensacola Mountains and the Forrestal Range (Himmelberg & Ford, 1975, 1976, 1983; Ford, 1976; Ford & Himmelberg, 1991). As already briefly mentioned, it consists overall of two major stratigraphic sections, thought to be a lower mafic part (Dufek, at ~3.5 km) and an upper more felsic part (Forrestal, at ~ 1.5 km). The Dufek pyroxenite layers contain strikingly similar Opx crystals, both in abundance, size, and character, right down to large (5–7 cm) annealed Opx crystal clots of uniform optical orientation, and with similar patterns of layering and anorthosite formation (Himmelberg & Ford, 1976). As surprising as it may be, in spite of a spade of detailed reports on the general geology, petrography, pyroxene and Fe–Ti oxide mineralogy, and isotope chemistry, there is a severe paucity of published whole rock analyses available to characterize the overall system, which greatly limits any significant petrogenetic comparison with the Dais Intrusion. Although outwardly the Dufek Intrusion is somewhat petrographically akin to the Dais Intrusion, but perhaps only on a more massive scale, its overall nature is distinct from the Dais in its great volumetric preponderance of gabbro with much smaller inter-layers of anorthosite and pyroxenite, capped by a massive (~300 m) granophyre (Ford, 1976; Ford & Himmelberg, 1991). Judging from the extensive compilation of bulk rock densities (~600 determinations), the volume of pyroxenite is relatively minor (Ford & Nelson, 1972). This overall compositional nature in terms of CaO–MgO is reflected in some representative whole rock analyses (Figure 12.1) (S. B. Mukasa, 2020, personal communication), where the Ferrar Dry Valleys compositional field is also shown. As previously discussed, the more mafic part of the Dry Valleys system is dominated by pyroxene fractionation, giving the characteristic tight correlation between CaO and MgO, similar to what is also found for olivine at Hawaii (Marsh, 2004, 2013). Although the Dufek data are sparse, there is much less indication of a prime role of pyroxene in controlling a fractionation trend, which itself is weak. The more fundamental meaning of these characteristics will be further discussed in Section 12.2.2.

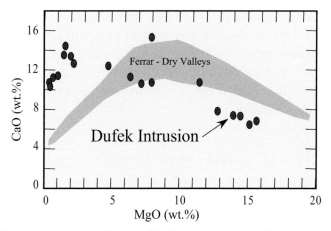

Figure 12.1 A comparison of the Dufek Intrusion and the Ferrar dolerites of the McMurdo Dry Valleys using a plot of whole rock CaO against MgO (wt.%). Although information for the Dufek is limited, it is much less cohesive and scattered relative to the dolerites of the Dry Valleys.

### *12.2.2 Central Transantarctic Region*

Between these two magmatic centers, Dufek and Dry Valleys, there is a fairly continuous distribution of dolerite sills throughout the Transantarctic Range, along with a smattering of associated lavas of Kirkpatrick basalt (e.g., Elliot & Fleming, 2017). Ice cover and lack of deep erosion in most areas prevents more comprehensive examination of the subjacent crust, but the general character is broadly similar to the upper sills in the Dry Valleys with a continued paucity of dikes. In an initial search for the telltale existence of large Opx in these sills that might reveal a nearby magmatic center, I have examined the extensive collection of dolerites established by the New Zealand Geological Survey housed at Lower Hut (N. Mortimer, pers. communication, 2009). This collection covers the region mostly from the Mesa Range, ~400 km north of the Dry Valleys, to the Beardmore Glacier area, about 800 km south, and in all of these specimens only two show clear signs of large pyroxene crystals. These are samples P21869 and P27518, from, respectively, Escalade Nunatek (collected by B. M. Gunn and G. Warren in 1958; also, Gunn, 1963) in the Boomerang Range about 150 km south of the Dry Valleys and Mt. Egerton in the Churchill Mountains about 350 km south of the Dry Valleys (collected by D. N. B. Skinner in 1961). Sample P21869 is a coarse-grained cumulate consisting mainly of clinopyroxene in plagioclase, whereas P27518 is a coarse cumulate of clinopyroxene and orthopyroxene quite similar to the Opx Tongue rocks, except being richer in Cpx. Also, near the Boomerang Range, in the northern part of the Warren Range, Grapes and Reid (1971) in a brief report show examples of strong rhythmic layering due to cumulate Opx and Cpx; a promised more thorough report is apparently yet to appear. Along these same lines, perhaps the most tantalizing field relations are in the region just north of the Beardmore Glacier, as reported by Grindley (1963, pp. 339–340):

Dolerite sills, dykes, plugs, and laccoliths are common in the Beardmore region. The plugs and laccoliths are found along and slightly east of the Plateau edge from the upper Marsh Glacier to the head of the Mill Glacier, and are considered to mark eruptive centres. The most important of these was found in the Marshall Mountains and may have been an active volcanic centre during Triassic times. Outside the eruptive centres, the Ferrar Dolerites occur as thick sills, commonly semi-concordant with the sediments but transgressive in detail, and connected by feeder dykes of all shapes and sizes. The sills range in thickness from as little as 50 ft to as much as 1,500 ft, with the usual thickness between 200 and 600 ft. Seven major sills totalling [sic] more than 3,000 ft in thickness, as well as numerous minor apophyses, were seen on the north face of Mt Mackellar.

Moreover, the petrography of the thicker sills is highly reminiscent of the early descriptions of the sills in the Dry Valleys:

The thicker sills are strongly differentiated, with accumulative hypersthene [i.e., Opx primocrysts] rich layers up to 150 ft in thickness in a 650 ft sill. The upper parts of some sills are marked by schlieren and intrusive veins of light-coloured felsic rocks of granophyric composition, representing the complementary differentiate. An account of the differentiation of similar dolerite sills in the McMurdo Sound area has been given by Gunn (1962), and differentiation of similar dolerite sills and dykes of approximately the same age in Tasmania is described by McDougall (1962).

This region seemingly has all the earmarks of a major intrusive center and this is also the type locality of the Kirkpatrick Basalts where a ~1,000 m section on Blizzard Peak in the Marshall Mountains, Queen Alexandra Range, exhibits 10 separate flows from 30 to 120 m in thickness where:

The thicker flows are visibly crystalline, resembling fine-grained dolerites, and like the dolerites are strongly differentiated. The lower parts of these flows contain accumulated hypersthene-rich layers produced by gravity differentiation. The accumulative layers are less than 30 ft thick and are not accompanied by complementary light-coloured felsic layers in the upper part of the flow, as in the thick differentiated dolerite sills of the area. Possibly a partly crystalline and differentiated magma was extruded from an intrusion undergoing gravity differentiation at depth. *(Grindley, 1963, p. 338)*

Subsequent investigations of sills in this area have not revealed anything of the strong ultramafic types found in the Dry Valleys. Nevertheless, one of the three sills at Mt. Achernar, which is about 50 km northwesterly of the upper Beardmore Glacier, between the Queen Elizabeth and Queen Alexandra Ranges, contains nearly 11 wt.% MgO at 21 m above the base (Faure & Mensing, 2010, p. 435). Altogether there are three sills here of 58, 126, and 67 m thickness, and it is the upper sill, oddly enough, that is enriched in magnesia; the others are of low

(~5–6 wt.%) magnesia. In this same general area, Portal Peak, about 50 km north of Mt. Achernar on the western edge of the Queen Alexandra Range, contains a 129 m thick sill (Barrett et al., 1986), which was analyzed in detail by Hergt et al. (1989a) and is remarkably uniform in magnesia, at 6–7 wt.%.

### 12.2.3 Beardmore to McMurdo

In between the Beardmore and the Dry Valleys are the Opx-rich and overlying Opx-poor sills at Roadend Nunatak; about 275 km south of the Dry Valleys. In terms of stratigraphic appearance and overall character, they have been linked to the Basement Sill and Peneplain Sill by Faure and Mensing (2010; p. 430). However, although possible, it is doubtful that these sills could be contiguous from the Dry Valleys occurrences. The Opx concentration is too large, relative to the general decline in Opx outward from Bull Pass and to what is found at Cathedral Rocks. This important discovery, instead, may indicate another major emplacement center in this general area.

In an effort to identify the mode of regional transport and emplacement of the Ferrar magmas, Elliot and Fleming (2017) give an insightful and valuable review of the occurrence of sills and associated lavas from the Mesa Range, north of the Dry Valleys, to the Beardmore Glacier region, some 800 km south of the Dry Valleys along the Transantarctic Range. Although they do not specifically mention the area described by Grindley, they list the MgO content of the chilled margins of a host of sills. Due to flow differentiation and the general mechanics of emplacement, however, chilled margins are notoriously unreliable indicators of the true mafic nature of a sill. The Basement Sill, for example, over an area of thousands of square kilometers has a chilled margin MgO content of ~7.25 wt.% when the nearby interior MgO content is commonly ~20 wt.%; this chilled margin constancy is true for sills worldwide.

Nevertheless, there is every indication that the sills in the Transantarctic Mountains are mainly high-level occurrences and are unlikely to carry high concentrations of mafic primocrysts. And being also essentially nonmagnetic, deeper-seated major intrusions will be difficult to find in aerial surveys, but they certainly may exist. In strong contrast to ocean ridge magmatism, where diking is commonplace, as exemplified by the major sheeted dike complexes found in ophiolites, and the major continental dike swarms associated with large mafic magmatic centers like Muskox, the Ferrar system is remarkably free of any preponderances of dikes, either as swarms or as obvious precursors to building a major central magmatic center. Yet the early Kirkpatrick lavas must have issued from fissure type systems, which are now inconspicuous (e.g., Elliot & Fleming, 2017), although the area described by Grindley may be a critical exception.

This paucity of major dike systems, coupled with the pervasive preponderance of sills throughout the Transantarctic Range along with telltale vestiges of cumulate Opx material here and there, clearly points to a series of undiscovered centralized magma systems similar to the Dufek and the Dry Valleys that gave rise to both the sills and lavas. The major piece of evidence linking the Ferrar sills and lavas to the deeper magmatic system is the voluminous Opx primocrysts. Where they came from and how they became so fully ingested into the Ferrar basaltic magma are key characteristics of this magmatic system.

# 13

# The Ferrar Magmatic Conundrum

## 13.1 Introduction

In terms of understanding the overall development of the Ferrar Province, perhaps the most curious characteristic of all is the extreme paucity of massive dikes, dike swarms, and the absence of well-developed rift valleys. The older, pre-Ferrar crustal rocks almost everywhere contain great numbers in length and width of early Paleozoic dikes, but none of these are Ferrar dikes. Wright Valley and Mt. Orestes areas, for example, are literally laced with dikes of many types. There are also certainly numerous small dikes associated with the advancing emplacement of the sills themselves, but there are none that can be specifically pointed to as massive fissure-style features able to carry the early magma through the crust. This is not a matter of exposure, as in the Dry Valleys the exposures are exceptional in three dimensions over perhaps ~50,000 km$^2$.

This extreme dearth of Ferrar dikes can only mean that the primary means of magma transfer was much more passive, more akin to a style of diapirism similar to what operates at deeper levels beneath ocean ridges. Should this be the case, a further enigma appears in that the apparent time to establish a passive system of magma transfer is expected to be long, relative to fissure and dike magma transfer; needing sufficient time to heat up the crust and ascent paths (e.g., Marsh, 1982; Rubin, 1991). But the entire Ferrar magmatic event was apparently remarkably short-lived (~0.4 Ma; Burgess et al., 2015), at least for the magmatism recorded in the Transantarctic Mountains. Moreover, what distinguishes the Ferrar system from other systems is that it is so relatively clean and isolated. Globally preserved dolerite-rift systems are generally due to a sudden jump of the developing rifts to another location, thereby isolating the already produced nascent rift system. These sudden jumps have clearly come at various times during development of the various fossil rifts (e.g., Brune et al., 2014).

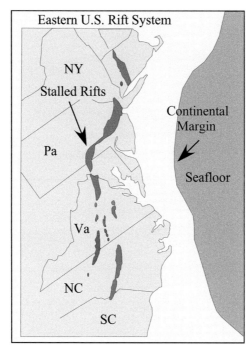

Figure 13.1 The dolerite rift basins of Eastern North America are contemporary with the Ferrar system, with many similar magmatic features

## 13.2 Other Gondwana Rift Systems

### 13.2.1 Eastern North America

In Eastern North America, for example, rifting produced a series of trough-like basins. Massive dolerites, in volume and number, are found primarily in North Carolina, Virginia, Pennsylvania, and New Jersey (Figure 13.1). A notable feature of these basins is the sequence of graben formation and dolerite emplacement. That is, these basins, like that at Gettysburg, are generally characterized on the west by high angle, easterly-facing listric-style normal faults, making the basins half grabens, hinged on the east (e.g., Thompson, 1998; Figure 13.2). The Gettysburg basin is filled with 5–9 km of fluvial and lacustrine sediments into which the dolerites have intruded, many reaching high into the sedimentary package prior to full sediment consolidation. The dolerite sills reflect this high-level emplacement by often being saucer-shaped and having irregular, locally bulbous contacts with patchy quenched chilled contacts appearing as wrinkled elephant skin. The dolerites clearly came late in the rifting, after the graben was already well developed and filled with large masses of sediment eroded from the nearby Appalachian Mountains. The patchy, although well developed, style of local basin development, as opposed to a single long sinuous rift, may indicate something

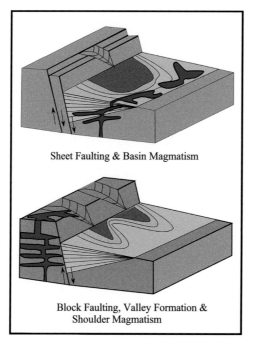

Figure 13.2 There is a strong contrast in style of rift development and timing of dolerite emplacement between the North American (upper) and McMurdo systems (lower)

about why it was aborted. And from the nature of the fluvial and lacustrine sediments, rift development was evidently slow as in the style of typical Atlantic opening. In the Hartford Basin, Philpotts (1992; Philpotts & Martello, 1986; Philpotts & Asher, 1993) has pieced together a detailed history of rift development based on the sequences of dike emplacement. Although the early rifting may have been as rapid as 90 mm/yr. in the widest part of the basin, in the more pervasive narrower southern part it started at ~65 mm/yr, but soon slowed to give an average over the whole event of ~40 mm/yr, and perhaps less. The associated dolerites and lavas do contain orthopyroxene phenocrysts, often in a reaction relationship with the melt, precipitating olivine upon ascent, except for the Palisades Sill, and the estimated near-surface temperatures at ~1,200°C (e.g., Philpotts, 1992) are similar to those of the Basement Sill.

For the Ferrar the timing and overall magma-tectonic coordination of development contrasts greatly to the more typical Gondwana rift development. First, there is little obvious sign that there was much of a typical massive rift system at all. There are no deep, large, sediment filled grabens with associated early dike swarms and lavas leading eventually to sill emplacement. The spatial scale of the size of the DV system, in length, breadth, and basin development is

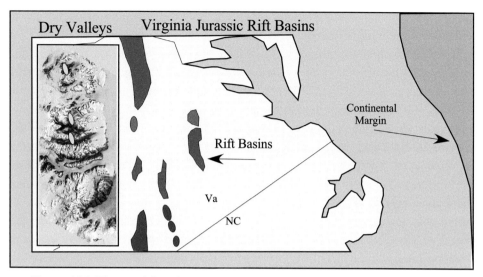

Figure 13.3 The spatial scale and possible basin development of the McMurdo system (left) is significantly smaller and more confined than that of the Eastern North American rift system

much less than the North American system; the DV system is spatial on the scale of the system found in Virginia (Figure 13.3). The thick Beacon Supergroup as a tectonic marker is mainly intact and dips gently westward, in apparent rebound, as expected for a sudden lithospheric rupturing (e.g., England, 1983; Buiter et al., 2009; Brune et al., 2014). The widespread pre-Ferrar Early Jurassic Hanson Formation records in inter-bedded sandstones and tufaceous sandstones distal silicic volcanism and apparently incipient rifting (Elliot, 1996; Elliot & Hanson, 2001; Elliot et al., 2016), but this may be due to sill emplacement. Much of this difference in style must be related to the significant contrast in the nature of the pre-existing continental crust. In Eastern North America, the nearby Appalachian Range represents the effects of an earlier continent–continent collision, with the resulting crustal structure containing the scars of this major suture (e.g., Puffer & Philpotts, 1988). The subsequent Mesozoic rifting reactivated much of this earlier rift, quite unlike the Ferrar system development, at least in the Dry Valleys region.

Second, dolerite sill emplacement came early, with apparently remarkably little wholesale disruption of the crustal structure, and these early intrusive magmas are at the heart of development of the entire system. At major points along the TAM highly localized magmatic centers evidently formed, spawning high-level extensive sheets, sporadically breaking open to the surface as tight clusters of locally intense diatremes, coalescing into a low-profile volcanic crater complex. Vigorous phreatic eruptions produced extensive tephra, characterized by the

Mawson Formation, leading to local dikes and more massive plug-like vents and to effusion of the Kirkpatrick Basalts.

Third, the local thick piles or welts of lava, in essence, progressively capped the system, with further extensive subsurface development, forming a stack of massive sills propagating downward, extending perhaps through the entire crust. In essence, instead of the intrusives arriving late, following extensive diking, producing distinctive telltale cross-cutting relationships, this sequence was the reverse, with the end result one of hiding all the early evolutionary evidence under local welts of tephra and lava. That is, because of this distinct reverse style of development, the ensuing eruptive products systematically covered the sequential evidential history of the magmatic–volcanic progression, making it all but impossible to unravel. Fundamentally key locations recording these events are found at the Coombs Hills and Allan Hills, and neighboring locals, in South Victoria Land as recognized by a number of early efforts, especially Grapes et al. (1974) and Korsch (1984), followed by extensive insightful investigations largely by James White and associates, principally Pierre-Simon Ross and Murray McClintock (e.g., Ross et al., 2008, and extensive references therein).

### 13.2.2 Coombs–Allan Hills Area

The Coombs–Allan Hills area is ~100 km N–NW of Bull Pass and exhibits a near paleo-surface vent regime exhibiting the intimate details of the transition from dolerite sill magma to diatremes, dikes, plugs, and lavas. Reorganization of the surface topography by doming and cracking, forming ridges and valleys, is the earliest signs of dolerite approaching the surface. From these heightened surfaces, massive lahar avalanches launched, remobilizing tephra, quenched dolerite, and blocks of Beacon sediments, filling valleys and depressions at highly variable levels. With the high temperature dolerite interacting with a high-level ground-water system, the vent area became an explosive phreato-magmatic cauldron spawning a tight cluster of energetic jets, eventually coalescing – presumably due to the cauterization sealing of the wet crater walls by quenched dolerite – into an insulated deep magmatic soup of blocks of Beacon immersed in dolerite. Thick (100s of meters) capping welts of dolerite lava (Kirkpatrick Basalt) issued mainly from plug-like conduits within the crater area and not from regional dikes. Large blocks of dolerite quenched around blocks of Beacon show signs of vesiculation, which is a clear indication of the initial magma water content, suggesting at this high level (i.e., low pressures) perhaps ~0.25 wt.%, similar to that of MORB or Hawaiian magma (e.g., Marsh & Coleman, 2009). Some of the lavas at nearby Carapace Nunatak show distinct vestiges of orthopyroxene phenocrysts, reminiscent of the Opx in the Basement Sill (Ross et al., 2008, see their figure 10e). Finally, judging from the vigor of the Dry Valleys magmatic system, there is every reason to expect that volcanic centers like this existed in many locations associated with the major magmatic centers.

### *13.2.3 Summary*

The unmistakable record is thus one of a long, linear, and narrow magmatic system rapidly initiated by the surreptitious migration of dolerite magma to the highest levels in the crust, reconfiguring the surface topography, explosively interacting with wet sediments, and producing a vast assemblage of volcaniclastics and lavas. There are some signs of rifting (Elliot, 2013), but it is not on a scale expected for a major Gondwana breakup as was so well established along Eastern North America (Figure 13.3). It may be more akin to local foundering of the crust in response to being literally floated over large areas by underlying thick, hot, low viscosity sheets of doleritic magma.

The exact origin and timing of the impressive frontal scarp (Figure 2.3) is not entirely clear (see Elliot, 2013, for a thorough review). David and Priestley (1914) considered this to be the Horst half of a typical Horst and Graben structure (Figure 2.27). And in many ways its origin is as similarly mysterious as that of, for example, the Sierra Nevada, where the sudden uplift came some 50 to 100 Ma after its major formative period of volcanism and plutonism (e.g., Huber, 1981; Henry, 2009; Mix et al., 2016), with a similar multitude of cogent arguments. The best that can be said is that the Ferrar LIP (Large Igneous Province) developed rapidly in the soon abandoned shoulder of what became a major oceanic rift, and that this region, similar to other Triassic–Jurassic rifts, took place along a continental margin where subduction, rifting, and arc accretion had gone on repeatedly since the Archean. That is, tectonic initiation was guided by a well-developed preexisting lithospheric structure. The surprising fact is the relative dynamic and geochemical coherence over such great distances (1,000s of kilometers) of the dominant quartz-normative doleritic magma. Moreover, the prevailing magma is not MORB-like tholeiite, showing clear signs of a peridotitic parentage after about 10–15 percent partial melting (e.g., Presnall & Hoover, 1987), but is something altogether quite different, much richer in silica and other crustal-like components. Yet, regardless of the exact tectonic evolutionary details, the Ferrar has broad petrologic affinities to other Mesozoic rift systems, like Eastern North America (e.g., Puffer & Philpotts, 1988). The layout of the system has all the hallmarks of sudden deployment followed by rapid isolation.

### 13.3 Origin Scenarios

The obvious scenarios of the origin and emplacement of doleritic magma of this nature are: (1) generation and magma dispersal from a massive dike swarm associated with a central plume; (2) more local magma production at mantle depths in an incipient lithospheric rift and dispersed as sills over substantial distances before escaping upward at specific locations; and (3) magma production everywhere along a mantle "rift" with magma migrating locally directly upward through the crust along the entire system. There are difficulties with each of these. First, there is no sign of any regional dike swarm, linear or radiating. Radiating plume-related dikes of this sort with widths of ~30 m can travel for well over 1,500 km, as exemplified by the Mackenzie dike swarm. AMS (Anisotropic Magnetic Susceptibility) and other flow indicators indicate that the flow begins nearly vertical and becomes increasingly horizontal with distance (Ernst, 1990; Ernst & Baragar, 1992). Although there is the semblance of a radiating dike swarm pattern in the early to middle Jurassic dolerite dikes from Western Dronning Maud Land (Riley et al., 2006) and in the Karroo system (e.g., Ernst & Buchan, 1998; Elburg & Goldberg, 2000; Aubourg et al., 2008; Sushchevskaya et al., 2009), nothing like this is seen throughout the Ferrar LIP; the Dronning Maud Land magmas also have a distinctly picritic flavor. Moreover, these magmas have clear MORB-like characteristics with $^{87}Sr/^{86}Sr_i$ ~0.7035 and $\varepsilon Nd_i$ ~9, with flat REE patterns in spite of high titania, quite unlike Ferrar magmas.

Regional radial dike patterns seem to be uncommon in typical Gondwana rift systems. The Eastern North America system, for example, shows plenty of dikes, but they are mainly focused along the rift structures themselves (e.g., Cummins et al., 1992) and, rather than forming a distinct radiating swarm, they reflect the prevailing regional stress field of the time. The complete paucity of major dikes throughout the Ferrar system suggests that, if dikes are present, they are highly focused and are hidden at some possibly subcrustal depth.

Magmatic dispersal for long distances in any kind of sheet – sills or dikes – suffers from the danger of magma heat death syndrome. That is, with cooling during transport, marginal solidification fronts migrate progressively inward with time, regardless of the rate of flow, which is orthogonal to the loss of heat. The net result is that the magma cannot travel any farther than the time taken for solidification[1] (e.g., Marsh & Coleman, 2009; Wright & Marsh, 2016). That is, for a sheet-like body of thickness H, the solidification time is $t \sim (H/2)^2/K$, where K is the thermal diffusivity ($\sim 10^{-2}$ cm$^2$/sec), and the maximum transport distance – ignoring thinning of the body due to geometric spreading during transport – is given by $L \sim Vt = V(H/2)^2/K$ or the necessary velocity to reach a distance L is $\sim K L/(H/2)^2$. For a 200 m thick sill, for example, the velocity of transport to reach 1,000 km must be at least ~10 m per day or about half a year to go 1,500 km.

---

[1] This concept is similar to realizing that the maximum distance a bullet can travel is given by the product of the velocity of the propelled bullet and time taken for the bullet to hit the ground when dropped by hand from the height of the gun. That is, these are two orthogonal vectors and cannot influence one another.

In and of itself, this is not unreasonable, but the magma must still get through the crust to erupt or be emplaced as a sill, and this does not consider any geometric thinning of the sill as it propagates outward. To prevent this and still travel 1,000s of kilometers and produce a narrow band of magmatism, the magma would need to be focused in a tight subcrustal rift or channel. And the transport dynamics do not include the requisite velocity to entrain, sort, and emplace high-level voluminous Opx-rich slurries. These constraints are greatly reduced if magma is produced at depth everywhere along a deep lithospheric rift, as along ocean ridges, and migrates to the surface locally by the combined processes of diking and diapirism.

The chief objection to this latter process is the recognition by D. H. Elliot and associates of a highly restricted distinctive fractionated composition (i.e., their SPCT type) of the youngest Kirkpatrick Basalt lava consistently found at various locations along the TAM from North Victoria Land to the Theron Mtns., covering some 3,000 km. This unusual distinction is attributed to derivation from a single large batch of magma transported via crustal dike swarms from a plume-like source along the TAM, with the magma breaking out to the surface at widely separated localities as governed by local rift-like structural controls (e.g., Elliot et al., 1999; Elliot & Fleming, 2017). And because this distinctive latest magma represents less than 1 percent of the volume of the more typical Ferrar magmas, but occupies the same plumbing as the dominant Ferrar magmas, it is inferred then that the whole Ferrar system might operate in this fashion of long transport from a central location near the African–Antarctic Jurassic juncture. The degree of primitiveness of the magma might be expected to diminish with distance from the origin (e.g., Leat, 2008). The overall long, narrow Ferrar belt reflects the focusing of this giant dike swarm at depth by an early Jurassic failed rift or a pre-existing lithospheric fabric parallel to the continental margin.

Impressed by the absence of any obvious well-developed surficial rift system and the seemingly neutral stress field reflected by local dikes near Allan Hills-style eruptive centers, Muirhead et al. (2012, 2014) instead suggest that the entire upper crust (~4 km), principally the Beacon section, becomes decoupled from the lower crust due to widespread sill emplacement. Once the dike-transported magma reaches the Beacon level in the crust, further transmission is via interconnected ramping and fault-connected sills, with central magmatic centers established at various locations to further disperse the radial emplacement of sill systems, such as is exemplified by the Dry Valleys system centered in Bull Pass. The vast nature of sills like the Basement Sill does clearly ensure that the Beacon crust over substantial areas was, indeed, floating. This scheme for sill emplacement and operation in the upper crust, as pertinent as it may be, does not help identify the ultimate source of the standard Ferrar basaltic magma. A valuable and thorough review of eruptive styles in these environments is given by White et al. (2009).

# 14

# Ferrar Magma Source Material

## 14.1 Plume Magmas

The conundrum of Ferrar magma emplacement over some 3,500 km has mainly centered on magma production and dispersal from a major plume involved in the initial Gondwana breakup. There is the impression that the compositional coherence of the Ferrar magmas demands a source where a large volume of semi-homogeneous magma can be readily produced and distributed across a long, narrow zone. Plumes can do this (Ernst & Buchan, 1998), but more typically the dike swarms are radial, and the magmas produced are not much like Ferrar magmas. Plume magmas are characterized by clear peridotitic affinities, which are commonly MORB-like with strong picritic overtones (e.g., Kogiso, 2007), more similar to the Dronning Maud Land dikes of Riley et al. (2006). This is nothing like the Ferrar, which has a much higher silica activity (i.e., siliceous with ~55 wt.% $SiO_2$), showing, at best, a long and tortuous fractionation path back to peridotite, with no sign of olivine domination in the actual rocks. Moreover, as is readily apparent from the McMurdo Dry Valleys sills, the magmas are also not especially homogeneous. Even intimately contiguous magmas filling the Peneplain Sill, for example, are in strong isotopic disequilibrium. On the other hand, plume-generated magmas having traveled some 2,500 km and then, having worked their way upward through the crust to the surface, could hardly be expected to be homogenous, even if they began life as homogeneous, which is unlikely. But the Ferrar inhomogeneity is distinctive and is on a much tighter spatial scale, suggesting that this characteristic was inherited much more locally, as at the base of the local subcontinental mantle. Plume-style magma dispersal is also exceedingly wasteful in that a great volume of magma is dispensed mostly to no observational avail, and tight regional dispersal is unlikely. It is also anomalous that a magmatic center as voluminous as the Dry Valleys would be produced so far from the point of initial generation near the Antarctic–Africa triple point. Furthermore, as discussed in Chapter 10, the extreme sub-chondritic Os/Ir ratios

($<0.33$), regardless of rock type (Dais Intrusion to normal Basement Sill), are quite unlike typical plume-derived magmas where Os/Ir = ~2.0, leading Choi et al. (2019) to suggest that these low Os/Ir values are instead more consistent with simple rift-style decompression melting.

## 14.2 Source Material Characteristics

In this regard, instead of a plume origin, derivation of the magmas from a linear source in the subjacent lithospheric mantle, enriched to varying degrees by a long history of subduction related processes, has also been readily proposed (e.g., Kyle et al., 1983; Cox, 1988; Hergt et al., 1989b, 1991; Molzahn et al., 1996; Hergt & Brauns, 2001, among others). This is usually discounted, however, by the assumption that the subcontinental lithosphere so produced will be highly heterogeneous and could never yield the Ferrar compositional coherence. The very same might be said for mantle plume and ocean ridge source materials and their derived magmas. For some additional perspective, island arc magmatism yielding high voluminous alumina basalt is characteristically compositionally highly coherent, including isotopically, over 1,000s of kilometers and, broadly speaking, even globally from one arc to another (e.g., Marsh, 2015). This subcrustal rifting process also suffers, however, from the necessity of having to produce high silica activity magmas from lithospheric, peridotitic source material. This problem is evaded by introducing by subduction a few percent of a continent-derived shale into a depleted lithospheric source, which may also produce more crustal-like trace elements (e.g., Hergt et al., 1989a, 1989b) along with hydrating or fluxing the source rock so that, once started, magma production is fast and voluminous (Choi et al., 2019). A small dose of sediment into depleted peridotitic source rock will help the trace element and isotopic constraints, but it is unlikely to yield a requisite source rock bulk composition to match the bulk composition of the common doleritic magmas so derived. And, as also discussed in Section 13.2, there is no sign whatsoever of any undue effect of water in the Ferrar dolerites, perhaps containing ~0.25 wt.% water.[1] The phreatic eruptions producing the Mawson Formation tephras, for example, are clearly due to high temperature magma interacting with adjacent high-level groundwater systems (e.g., White & McClintock, 2001).

Beyond the marked defining characteristics of having high silica, relatively low concentrations of alkalis, magnesia, titania, and $P_2O_5$, and greatly enriched initial Sr and Nd isotope ratios, the common Ferrar Carrier Magmas show a strong crustal

---

[1] Some minor signs of dolerite vesiculation near the contacts with the Beacon sediments have been noted by Ferrar (1907) and Smith (1924), but these are probably due to local boiling of groundwater.

trace element signature, which is closely correlated with the *local* upper crust (e.g., Mensing et al., 1984; Hergt et al., 1989a; Fleming et al., 1995, 1997; Molzahn et al., 1996; Brauns et al., 2000). As has already been shown in some detail for the magmatism in Bull Pass, with the extensive melting of the granitic crust by the dolerites, including extensive networks of re-injected aplitic dikes, there is ample opportunity for contamination (see Figures 3.2, 6.7, and 6.8), but there is little sign in the dolerites themselves of this process actually taking place at this high level in the crust. That is, the sills and the silicic melts are spatially distinct with no signs whatsoever of any wholesale intimate mixing or mingling taking place during the emplacement process, even though there are obvious implied dynamic processes that might carry out intimate mixing (Hersum et al., 2007; see Figure 6.8). Judging from the sill chemical stratigraphy, keeping in mind that these sills are all multiply injected, it seems more likely that there are separate significant volumes of magma that may be much more homogeneous than the entire sill as a whole. Taken altogether this suggests that the distinctive siliceous bulk composition of the Carrier Magma is being generated locally in significant volumes in the subcontinental mantle in a source rock that is, in and of itself, in strong contrast to peridotitic mantle, depleted and/or contaminated or not. And the relatively late entrainment of the Opx slurries, along with their isotopic contrast, suggests this magma production is taking place beneath the source of the Opx mass. Besides the systematic generation of these distinctive dolerites, a major indication of this refined source composition is the fact that great volumes of this magma can be produced rapidly upon simple rifting. This ease of rapid melt production may well strongly reflect generation from a basaltic or eclogitic source rock.

Interestingly enough, in a study of the anomalous oxygen isotope composition of Karoo and Etendeka picrites, Harris et al. (2015) conclude that the mantle source region itself is anomalous in its isotopic oxygen, that it has been like this possibly since Bushveld times, and that it may reflect a strong eclogitic component from emplacement of oceanic lithosphere into the cratonic keel during Archaean subduction.

## 14.3 Source Types and Melting Styles

Ocean ridge tholeiites, to first order, show a remarkable global homogeneity reflecting compositional buffering, both in terms of bulk magma and isotope composition, by a peridotitic source material. Adiabatic convective up flow systematically leads to partial melting and escape of melt after 10–15 vol.% melting. The regularity of this process, to a great extent, reflects the phase equilibria of peridotitic source rock (Figure 14.1), which at lower pressures has a large temperature range of about 600°C between solidus and liquidus (e.g.,

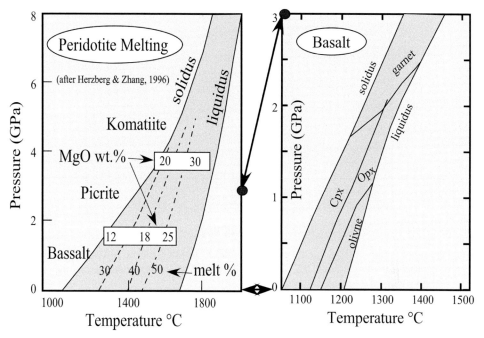

Figure 14.1 Phase equilibria in the peridotite and tholeiitic basalt systems. At relatively low pressures (~2 GPa) the full melting interval in the peridotite system is ~600°C and extends to ~1,700°C, which during adiabatic ascent greatly limits or buffers the overall degree of melting. This is greatly in contrast to the basalt system where the full melting range at 1.5 GPa is only about 100°C, which upon modest changes in temperature allows for voluminous melt production.

Herzberg & Zhang, 1996). This large temperature range prevents melting from becoming too extensive and, in essence, dynamically buffers the whole process of melt extraction and chemical equilibration to occur repeatedly at more or less the same degree of melting. Unusually large volumes of melt are prevented from forming because the temperature rise is greatly limited by the heat of fusion and the large-scale adiabatic convection itself. The fact that unusually large volumes of Ferrar basaltic magma were produced over a very limited time (<~0.4 Ma) with apparently relatively minor and limited tectonic disturbance, presumably subcrustal rifting, reflects magma generation from a distinctly different rock source.

The melting relations for a typical MORB-like tholeiitic bulk composition is strikingly different than for a peridotitic source rock (Figure 14.1). At moderate pressures (~1 GPa) the spread between solidus and liquidus is less than 100°C, and it may be even narrower at higher pressures (e.g., Cohen et al., 1967). This is a reflection mostly of the transition to a quartz eclogite where the phase assemblage changes from olivine, plagioclase, and pyroxene to plagioclase, pyroxene, quartz,

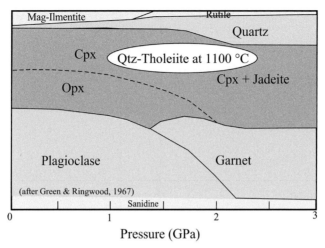

Pressure (GPa)

Figure 14.2 Sub-solidus modal proportions in a quartz tholeiite bulk composition as a function of pressure. At pressures up to about 2 GPa, orthopyroxene and clinopyroxene are dominant phases and with more MgO and less $SiO_2$ this field expands (after Green & Ringwood, 1967).

and garnet, with garnet becoming increasingly dominant with increasing pressure, especially after the loss of plagioclase. Olivine is not stable beyond about 0.8 GPa. Unlike for peridotitic material, this combination of an unusually narrow melting range and a simple phase assemblage permits large volumes of melt production with modest changes in temperature. It is likely that a mechanism like this is responsible for the voluminous outpouring of the typical Ferrar type Basaltic Carrier Magma, although I hasten to add that the exact bulk source composition is most probably not MORB, but something more distinctive. This phase behavior is common for all basaltic compositions, and the more detailed phase delineations for a quartz tholeiite (Figure 14.2) reveals a strong presence of orthopyroxene up to pressures of about 2 GPa, along with the presence of rutile and sanadine. The phase assemblage and the melting interval may be sensitive to the exact basalt bulk composition, but the melt composition (Figure 14.3), especially for $SiO_2$ and MgO, is almost invariant from 1 to 5 GPa (e.g., Pertermann & Hirschmann, 2003a, 2003b; Spandler et al., 2007; Marsh, 2015). In this system, the typical Ferrar Basaltic Carrier Magma bulk composition of ~55 wt.% $SiO_2$ and 4–6 wt.% MgO reflects degrees of melting of 40–50 vol.%. The almost universally low MgO and elevated $SiO_2$ contents of these voluminous magmas may be highly diagnostic of coming from a parent rock type similar to quartz eclogite, especially at moderately low pressure before garnet becomes a dominant residual phase to heavily influence the REE budget (e.g., Brophy & Marsh, 1986).

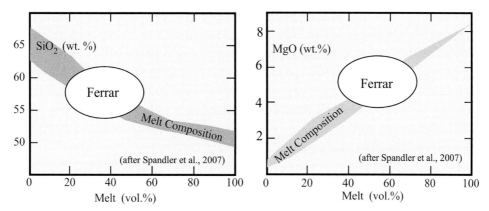

Figure 14.3 The variation in melt composition of $SiO_2$ and MgO (both wt.%) for melting of basaltic bulk compositions up to pressures of 5 GPa. The effect of pressure for these components is minimal and for typical Ferrar dolerites compositions the degree of melting might be ~50 vol.% or more.

## 14.4 Crust–Mantle Communication

Given the distinctive composition of the Basaltic Carrier Magma along with its affinity for local upper crustal trace elements, and the telltale evidence throughout the Ferrar LIP of pervasive Opx primocrysts, equilibrated at high temperature, points to a subcrustal stratified mantle characterized by orthopyroxene rather than olivine. The collective evidence points to fundamental ongoing processes of crust–mantle interaction operating systematically over long periods of time. In this regard, it has long been noted that the continental crust collectively receives a steady, albeit episodic, flux of basaltic magmatism from rifting, regional dike swarms, random cinder cone volcanism, and even massive flood basalts. Since there is no obvious ongoing process whereby continental crust is systematically purged of this basaltic component, the continental crust might be expected over time to become increasing basaltic. The fact that it apparently doesn't, but remains largely andesitic (e.g., Rudnick, 1995; Rudnick & Gao, 2005), suggests that there must be a process or processes whereby this component is systematically purged. The processes available to perform this purging are limited, being mainly differentiation due to erosion, density or local buoyancy contrasts, and internal magmatism.

Erosion breaks down the individual units into mineral components, some of which become clays and others become fully dissolved in aqueous solutions (e.g., cations such as $Na^{2+}$, $Mg^{2+}$, $Ca^{2+}$), travel back to the oceans, and eventually may be deposited via subduction back into the deeper mantle or pasted beneath continental margins. Other components, such as silica and alumina, remain in the crust. Basaltic units deeper in the crust may, due to density, under the appropriate conditions slowly migrate progressively deeper and transform into granulitic or eclogitic rock, thus becoming even denser. Once in the lower crust, these bodies may individually sink into the uppermost mantle, or perhaps collect in sufficient concentrations in the lowermost crust and cause large fragments of the lower crust to delaminate and enter the upper mantle. Many variations on this theme, loosely referred to as delamination, have been defined and promoted (e.g., Ringwood, 1975; Anderson, 2007), and this vast literature is well summarized by Ducea (2011), among others (e.g., Arndt & Goldstein, 1989).

The crust might also be reworked and differentiated by the prolonged interaction of island arc-related magmatism along continental margins, which has also been suggested for the Ferrar system. For this process, there have also been some remarkably thorough petrologic studies of the lower crust describing and quantitatively documenting the interaction between arc-derived basaltic magmas and continental crust (e.g., Walker et al., 2015). More specifically, this latter study of the Sierram Valle Fertil Complex of west–central Argentina represents a large crustal piece of the Ordovician (~470 Ma) Famatinian arc, which exposes a continuous, tilted crustal arc section ranging in depth from ~12 to 32 km. This insightful study and others like it suggest that the overall arc magmatic process is, by and large, chaotic, and reminiscent of prolonged and repeated crustal melting, assimilation, storage, and homogenization, which is essentially the so-called MASH process of Hildreth and Moorbath (1988). The in situ mafic magmas are, indeed, often gabbro-norites, broadly similar to the Dais magmas, but there is also a very broad spectrum of associated compositions, both much more silicic and more mafic. This is in strong contrast to the Ferrar LIP system where the magma type, although isotopically heterogeneous, is much narrower and well defined. The presence of large reserves of orthopyroxene distinguishes the Ferrar source region as distinctive from arc magmatism related processes that rework and differentiate the crust. And these characteristics are not special to the Ferrar system.

The Mesozoic Eastern North American Rift System exhibits abundant concentrations of orthopyroxene similar to that found in the Ferrar system (e.g., Gottfried & Froelich, 1985, 1988; Puffer & Philpotts, 1988; Cummings et al., 1992; Ragland et al., 1992; Mangan et al., 1993). Stretching from near York Haven, Pennsylvania, southward to the Gettysburg Basin and onwards south to the Culpepper basin of Virginia, are many locations where high concentrations of large orthopyroxene can be found in the dolerite sills. Particularly striking examples of this rock, which has been quarried for building stone, has been used to construct the Visitors Center at the Gettysburg National Park site and also the headquarters of the American Association for the Advancement of Science in Washington, DC. Excellent examples of Opx sorting and layering and schlieren of plagioclase are beautifully exposed throughout these buildings. Other examples where magmatic systems are dominated by massive amounts of primocrystic orthopyroxene are the Stillwater Complex of Montana, the Bushveld Intrusion of South Africa, and the Great Dike of East Africa. The former two bodies, being much older than the Mesozoic rift systems, may indicate that this process of crust–mantle mixing has been active for much of Earth history. The apparent fact that random rifting repeatedly exposes so broadly similar magmatic products implies that this is a global process and not simply restricted to certain continental margins. In this regard, it must also be mentioned that there are other areas that carry abundant slurries of primocrystic olivine and not orthopyroxene. Some of the more prominent of these areas are the Muskox Intrusion, the Eocene magmatic centers of Western Scotland, like Rum and Skye, and the ultramafic centers of southeastern Alaska, like that at Duke Island. It may be notable that these locations are largely on the edges of continents, whereas the orthopyroxene-dominated systems have much stronger intra-continental affinities.

## 14.5 Conclusions

This altogether suggests that the sub-continental mantle region spawning the dolerites typical of Gondwana breakup magmatism is commonly dominated by a much more siliceous (high silica activity) voluminous, pyroxene-rich ultramafic country rock. This implies a systematic, even global, process of crustal material locally re-entering the mantle through delamination or foundering of large pieces of eclogitic and/or granulitic rock. And as is typical of a body immersed in a highly viscous fluid (crust), in leaving the crust these basaltic blobs and dollops will carry with them thick coatings of silicic crustal material. By this process the subcontinental mantle receives a steady flux of crustal material, over time achieving a close chemical intimacy with the immediately overlying continental crust. The highly radiogenic and isotopically heterogeneous nature of the lavas, pure dolerites, and the orthopyroxene-rich rocks implies that this more siliceous mantle region is itself highly heterogeneous, but on a local or regional spatial scale, perhaps reflecting the ultimate origin of the individual foundered pieces of crust.

Finally, the enduring Ferrar geochemical conundrum may well have been similarly summed up by Brauns et al. (2000, p. 905) in their study of the Tasmanian dolerites: "Instead, the results are more readily explained by models in which the introduction of upper-crustal materials into the mantle contaminates the source of these magmas before melting."

Knowing this focuses additional attention on the more detailed structural and petrologic nature of this region. That this subcontinental mantle is, in and of itself, clearly layered is reflected in the ultimate deeper-seated source region of the dolerites themselves and by the entrainment of overlying pyroxene dominated

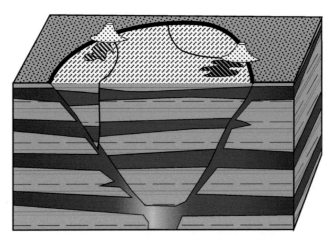

Figure 14.4 A schematic illustration of what the Ferrar magmatic system in the McMurdo Dry Valleys above Bull Pass may have resembled during the later stages of formation

ultramafic country-rock. This apparent stratigraphic arrangement suggests the dolerite source rock to be denser than the pyroxenite, indicating an eclogitic nature. The dolerite source material is also extensive and compositionally coherent, as reflected in its ability to readily produce broadly similar magma throughout the Ferrar LIP, which adheres to near MORB-like phase equilibria, but its strong isotopic and trace-element heterogeneity clearly reflects systematic regional variations indicative of its multi-varied origins. The pervasive characteristic dolerite bulk compositions and highly uneven volumetric outpourings further suggest, within the region as a whole, significant variations in source melting-characteristics.

The Ferrar magmatic system exemplified in the McMurdo Dry Valleys, which may be the largest in all of the Ferrar LIP, is one of the more revealing magmatic systems yet known. In its singular three-dimensional pristine exposures over a vast region (~10,000 km$^2$) it displays the intimate inner workings of a magmatic system bridging the critical juncture between plutonism and volcanism. It may well represent the head of an extensive magmatic mush column (Figure 1.1) revealing the full history of initial establishment and evolution as a stack of massive basaltic sills propagating outward from a central location in Bull Pass (Figure 14.4).

The system was built from the top down due to a combination of magmatic overpressure and deep and local density controls. The lowermost of the four major sills, the Basement Sill, contains a voluminous slurry of orthopyroxene-dominated primocrysts/xenocrysts whose spatial distribution provides a detailed record of the dynamics of emplacement and fractionation leading to differentiation, in addition to providing a clear and extensive sampling of the subcontinental mantle.

All major differentiation is due to mechanical fractionation processes involving the Opx primocrysts. The uppermost sills overlap in bulk composition with local flood basalts and the deepest ponding, some 4 km below, in the Dais Intrusion has many of the characteristics of a major ultramafic layered intrusion. Its relatively short solidification time (~2,000 years) has captured unusual processes of layer formation, crystal sorting via particle sieving, phenocryst production and growth via annealing, along with many other processes hitherto previously only hinted at and applicable to many other intrusions.

The entire system is also compositionally fairly cohesive in terms of bulk composition, adhering well to MORB-like CMAS phase equilibria, yet it is everywhere strongly isotopically heterogeneous, being characterized by two end members: A Basaltic Carrier Magma, producing all the dolerites and lavas, and the massive ultramafic slurry, the two of which are in strong compositional and isotopic disequilibrium. The dolerite magma is broadly similar to that found throughout the Ferrar LIP and it is the essential working fluid giving rise to the entire magmatic system. From the irregularly distributed voluminous regional outpourings, coupled with the rapid establishment of the Ferrar system, taking ~0.4 Ma, the magmatic source is inferred to possess a narrow melting range, perhaps similar to quartz eclogite, and lying beneath the ultramafic pyroxenite, which is likely immediately beneath the crust. That these features are globally common to other rift-related dolerite systems collectively points to long-term ongoing processes of exchange between the crust and mantle such that the immediate subcontinental mantle is dominated by pyroxene mineralogy rather than typical olivine dominated peridotite.

### *14.5.1 A Final Word*

I hasten to add that the present work only scratches the petrologic and geologic surface of the true nature of this marvelous magmatic system. In a very real way, we have found and explored some of the more obvious and fascinating aspects of this unusually well exposed system. It certainly may contain additional complexities, as in the region south of Finger Mountain toward the Lashly Mountains, and also to the north, which may be fundamentally important to understanding the broader basic magmatic architecture of Ferrar magmatism, especially revealing possible connections between adjacent magmatic centers along the Transantarctic Mountains. Bull Pass itself warrants (see Figure 14.5) a more exacting thorough examination to reveal and understand the many intimate unusual interactions between the basaltic magma and the highly remobilized granitic crust.

To emphasize a little of the geological value of this special location, Figure 14.6 shows two views of the south end of Bull Pass. One view (on the left) looks

Figure 14.5 Bull Pass magmatic center. Looking northwesterly across Wright Valley through Bull Pass and into McKelvey Valley. The image on the left is a New Zealand Geological Survey map projected onto a digital elevation surface, and the image on the right is the actual terrain where the outline of the Basement Sill can be seen in the valley walls.

Figure 14.6 Two views, looking east (left) and west (right), at the south end of Bull Pass, emphasizing the many features of the Basement Sill emplacement process exposed in great detail (see the text for more explanation)

eastward, where the lower contact of the Basement Sill is clear in the foreground. The sill trends off in the distance along the north wall of Wright Valley, where the presence of the Opx Tongue is delineated by the thick ribbon of sand within the sill. The Ribs area is in the upper left where, leading up to it, are quenched patches of the sill on the dip slope, and the thin sill-like leading protrusions, marking the cutting-edge during progressive emplacement, are in the foreground on the right.

The other view, looking westerly, shows the massive, full thickness of the sill as it trends up Wright Valley toward the Dais, the ragged texture of which indicates the presence of the very coarse, highly concentrated Opx Tongue and, at the top edge, the strong climbing contact. All along the lower contact, exposed just over the edge in Wright Valley, are local quenched dolerite and melted granite sites, intermingling on a scale of tens of centimeters. The sandy area at the mid-slope of the sills is comprised of dunes of Opx Tongue materials. Mt. Jason is in the distance at the top, which is the Peneplain Sill.

When the sill is traced northward up Bull Pass the inter-relations between these lobes become more intricate and demanding in terms of sorting out the sequence of events, giving the impression of perhaps being the roots of a once vigorous cauldron.

The rewards to be gained here promise to be highly educational and singularly valuable to understanding magmatic processes.

# References

Ablay, G. J., Clemens, J. D., Petford, N. (2008) Large-scale mechanics of fracture-mediated felsic magma intrusion driven by hydraulic inflation and buoyancy pumping. *Geological Society, London, Special Publications*, **302**, 3–29.

Alexander, C. (1998) *The Endurance: Shackleton's Legendary Antarctic Expedition*. Knopf Doubleday Publishing Group, New York, 224 p.

Allibone, A. H., Cox, S. C., Graham, I. J., Smellie, R. W., Johnstone, R. D., Ellery, S. G., Palmer, K. (1993a) Granitoids of the Dry Valleys Area, Southern Victoria Land, Antarctica: Plutons, field relationships, and isotopic dating. *New Zealand Journal of Geology and Geophysics*, **36**, 281–297.

Allibone, A. H., Cox, S. C., Smillie, R. W. (1993b) Granitoids of the Dry Valleys Area, Southern Victoria Land, Antarctica: Geochemistry and evolution along the early Paleozoic Antarctic Craton margin. *New Zealand Journal of Geology and Geophysics*, **36**, 299–316.

Allibone, A. H., Forsyth, P. J., Sewell, R. J., Turnbull, I. M., Bradshaw, M. A. (1991) Geology of the Thundergut area, southern Victoria Land, Antarctica. 1:50 000 miscellaneous series map 21 (with supplementary text). Geology and Geophysics Division. Department of Scientific and Industrial Research, Wellington, New Zealand.

Anderson, D. L. (2007) *New Theory of the Earth*. Cambridge University Press, Cambridge, 384 p.

Anderson, S. W., Stofan, E. R., Smrekar, S. E., Guest, J. E., Wood, B. (1999) Pulsed inflation of pahoehoe lava flows for flood basalt emplacement. *Earth and Planetary Science Letters*, **168**, 7–18. (and Reply to Self et al. (2000) discussion of above.)

Aranson, I. S., Tsimring, L. (2009) *Granular Patterns*. Oxford University Press, Oxford, 343 p.

Armitage, A. B. (1905) *Two Years in the Antarctic: Being a Narrative of the British National Antarctic Expedition*. Edward Arnold, London, 315 p.

Armstrong, R. L. (1978) K-Ar dating: Late Cenozoic McMurdo Volcanic Group and Dry Valley glacial history, Victoria Land, Antarctica. *New Zealand Journal of Geology and Geophysics*, **21**, 685–698.

Arndt, N. T., Goldstein, S. L. (1989) An open boundary between lower continental crust and mantle: Its role in crust formation and crustal recycling. *Tectonophysics*, **161**, 201–212.

Arndt, N. T., Guitreau, M., Boullier, A.-M., Le Roex, A., Tommasi, A., Cordier, P., Sobelev, A. (2010) Olivine, and the origin of kimberlite. *Journal of Petrology*, 51, 573–602.

Aubourg, C., Tshoso, G., Le Gall, B., Bertrand, H., Tiercelin, J. J., Kampunzu, A. B., Dyment, J., Modisi, M. (2008) Magma flow revealed by magnetic fabric in the Okavango giant dyke swarm, Karoo igneous province, northern Botswana. *Journal of Volcanology and Geothermal Research*, **170**, 247–261.

Bachmann, O., Bergantz, G. W. (2008) Rhyolites and their source mushes across tectonic stings. *Journal of Petrology*, **49**, 2277–2285.

Baragar, W. R. A. (1960) Petrology of basaltic rocks in part of the Labrador Trough. *Bulletin of the Geological Society of America*, **71**, 1589–1644.

Barrett, P. J. (1991) The Devonian to Triassic Beacon Supergroup of the Transantarctic Mountains and correlatives in other parts of Antarctica. In: Tingey, R. J. (ed.) *The Geology of Antarctica*. Oxford Monographs on Geology and Geophysics, **17**. Oxford University Press, Oxford, 120–152.

Barrett, P. J., Elliot, D. H., Lindsay, J. F. (1986) The Beacon Supergroup (Devonian–Triassic) and Ferrar Group (Jurassic) in the Beardmore Glacier area, Antarctica. In: Turner, M. D., Splettstoesser, J. F. (eds.) *Geology of the Central Transantarctic Mountains*. Antarctic Research Series, **36**. American Geophysical Union, Washington, DC, 339–428.

Batiza, R. (1996) Magmatic segmentation of mid-ocean ridges: A review. In: Macleod, C. J., Tyler, P. A., Walker, C. L. (eds.) *Tectonic, Magmatic, Hydrothermal and Biological Segmentation of Mid-Ocean Ridges*. Geological Society, London, 103–130.

Bedard, J. H. (2005) The roots of a flood basalt province: Expedition to Antarctica. *Elements*, December, 316–317.

Bedard, J. H., Marsh, B. D., Hersum, T. G., Naslund, H. R., Mukasa, S. B. (2007) Large-scale mechanical redistribution of orthopyroxene and plagioclase in the Basement sill, Ferrar Dolerites, McMurdo Dry Valleys, Antarctica: Petrological, mineral–chemical and field evidence for channelized movement of crystals and melt. *Journal of Petrology*, **48**, 2289–2326.

Benson, W. N. (1916) Report on the petrology of the dolerites collected by the British Antarctic Expedition, 1907–1909. British Antarctic Exped. 1907–9. Repts. *Science Investigations*, **2**(9), 153–160.

Bergantz, G. W. (1989) Underplating and partial melting: Implications for melt generation and extraction. *Science*, **245**, 1093–1095.

Bergantz, G. W., Schleicher, J. M., Burgisser, A. (2017) On the kinematics and dynamics of crystal-rich systems. *Journal of Geophysical Research: Solid Earth*, **122**, 6131–6159.

Bhattacharji, S. (1967) Mechanics of flow differentiation in ultramafic and mafic sills. *Journal of Geology*, **75**, 101–112.

Bickel, L. (2001) *Shackleton's Forgotten Men: The Untold Tragedy of the Endurance Epic*. Pimlico Press, London, 241 p.

Boudreau, A. E. (1995) Crystal aging and the formation of fine-scale igneous layering. *Contributions to Mineralogy and Petrology*, **54**, 55–69.

Boudreau, A. E. (2011) The evolution of texture and layering in layered intrusions. *International Geology Review*, **53**, 330–353.

Boudreau, A. E., McBirney, A. R. (1997) The Skaergaard layered series. Part III. Non-dynamic layering. *Journal of Petrology*, **38**, 1003–1020.

Boudreau, A. E., Simon, A. (2007) Crystallization and degassing in the basement sill, McMurdo Dry Valleys, Antarctica. *Journal of Petrology*, **48**, 1369–1386.

Bowen, N. L. (1915) The later stages of the evolution of the igneous rocks. *Journal of Geology*, **23**, 1–91.

Brauns, C. M., Hergt, J. M., Woodhead, J. D., Maas, R. (2000) Os isotopes and the origin of the Tasmanian dolerites. *Journal of Petrology*, **41**, 905–918.

Bray, B., Harrp, K. S., Geist, D., Garcia, M. O., Swarr, G. J., Garman, K. A., Buck, S. A., Parcheta, C. E., Matulattis, I. (2009) Sr and Nd isotopic and geochemical analysis of the Vanda Dike Swarm, Antarctica. *AGU Fall Meeting Abstracts*, V51C–1685..

Brey, G. P., Kohler, T. (1990) Geothermobarometry in four-phase lherzolites. II. New thermobarometers, and practical assessment of existing thermobarometers. *Journal of Petrology*, **31**, 1353–1378.

Brook, E. J., Brown, E. T., Kurz, M. D., Ackert, R. P., Raisbeck, G. M., Yiou, F. (1995) Constraints on age, erosion, and uplift of Neogene glacial deposits in the Transantarctic Mountains determined from in-situ cosmogenic $^{10}$Be and $^{26}$Al. *Geology*, **23**, 1063–1066.

Brophy, J. G., Marsh, B. D. (1986) On the origin of high-alumina basalts and the mechanics of melt extraction. *Journal of Petrology*, **27**, 763–789.

Browne, W. R. (1923) The dolerites of King George Land and Adelie Land. *Scientific Reports of the Australasian Antarctic Expedition 1911–1914*, 3, 245–258.

Brune, S., Christian Heine, C., Pérez-Gussinye, M., Sobolev, S. V. (2014) Rift migration explains continental margin asymmetry and crustal hyper-extension. *Nature Communications*, **5**, 4014.

Buiter, S. J. H., Pfiffner, A. O., Beaumont, C. (2009) Inversion of extensional sedimentary basins: A numerical evaluation of the localisation of shortening. *Earth and Planetary Science Letters*, **288**, 492–504.

Bull, C. (2009) *Innocents in the Dry Valleys: An Account of the Victoria University of Wellington Antarctic Expedition 1958–1959*. Victoria University Press, Wellington, 267 p.

Burgess, S. D., Bowering, S. A., Fleming, T. H., Elliot, D. H. (2015) High precision geochronology links the Ferrar Large Igneous Province with early Jurassic ocean anoxia and biotic crisis. *Earth and Planetary Science Letters*, **415**, 90–99.

Cartwright, J., Hansen, D. M. (2006) Magma transport through the crust via interconnected sill complexes. *Geology*, **34**(11), 929–932.

Cawthorn, R. G. (2002) Delayed accumulation of plagioclase in the Bushveld complex. *Mineralogical Magazine*, **66**, 881–893.

Chaloner, B., Kenrick, P. (2015) Did Captain Scott's Terra Nova expedition discover fossil nothofagus in Antarctica? *The Linnean*, **31**(2), 1–17.

Chaloner, B., Kenrick, P. (2016) Correction. *The Linnean*, **32**(1), 7–8.

Charrier, A. D. (2010) Emplacement history of the Basement Sill, Antarctica: Injection mechanics of crystal-laden slurries. PhD dissertation, Johns Hopkins University, Baltimore, MD, 300 p.

Charrier, A. D., Marsh, B. D. (2004) Sill emplacement dynamics from regional flow sorting of opx phenocrysts, Basement Sill, McMurdo Dry Valleys, Antarctica. American Geophysical Union Meeting, Montreal, abstract V42A-03.

Charrier, A. D., Marsh, B. D. (2005) Sill emplacement dynamics: Experimental textural modeling of a pulsing, cooling, particle-laden magma as applied to the Basement Sill, McMurdo Dry Valleys, Antarctica. American Geophysical Union Meeting, San Francisco, abstract V23A-0686.

Charrier. A. D., Marsh, B. D. (2021) Linking the pulses of volcanism with subsurface magma dynamics: Part 1. On the interaction of advancing solidification fronts and flow-sorted particle distributions. (to be submitted).

Choi, S. H., Mukasa, S. B., Ravizza, G., Fleming, T. H., Marsh, B. D., Bédard, J. H. J. (2019) Fossil subduction zone origin for magmas in the Ferrar Large Igneous

Province, Antarctica: Evidence from PGE and Os isotope systematics in the Basement Sill of the McMurdo Dry Valleys. *Earth and Planetary Science Letters*, **506**, 507–519.

Clague, D. A., Dalrymple, G. B. (1987) The Hawaiian-Emperor volcanic chain. Part I, geologic evolution. In: *Volcanism in Hawaii. U.S. Geological Survey Prof. Pap., 1350*, 5–54. Washington, DC.

Cohen, L. H., Ito, K., Kennedy, G. C. (1967) Melting and phase relations in an anhydrous basalt to 40 kilobars. *American Journal of Science*, **265**, 475–518.

Cooke, W., Warr, S., Huntley J. M., Ball, R. C. (1996) Particle size segregation in a two-dimensional bed undergoing vertical vibration. *Physical Review Letters*, **53**, 2812–2822.

Cotton, L. A. (1916) Petrographical notes on some rocks retrieved from the cache at Depot Island, Antarctica. Report of British Antarctic Expedition. 1907–09. *Geology*, **2**(13), 235–237.

Cox, K. G. (1988) The Karroo province. In: MacDougall, J. D. (ed.) *Continental Flood Basalts*. Kluwer, Boston, MA, 239–271.

Cox, S. C., Turnbull, I. M., Isaac, M. J., Townsend, D. B., Smith Lyttle, B. (compilers) (2012) Geology of southern Victoria Land, Antarctica. Institute of Geological & Nuclear Sciences 1:250 000 geological map 22. 1 sheet +135 p. Lower Hutt, New Zealand. GNS Science.

Cummins, L. E., Ragland, P. C., Arthur, J. D. (1992) Classification and tectonic implications of early Mesozoic magma types of the Circum-Atlantic. *Geological Society of America, Special Paper*, **268**, 119–135.

Currier, R. M. (2011) Three aspects of magma dynamics: Granite partial melting, experiments on laccolith emplacement, and effects of xenocryst re-equilibration. Doctoral Dissertation, Johns Hopkins University, Baltimore, MD, 220 p.

Currier, R. M., Flood, T. P. (2019) The Orestes melt zone, McMurdo Dry Valleys, Antarctica: Spatially distributed melting regimes in a contact melt zone, with implications for the formation of Rapakivi and Albite granites. *Journal of Petrology*, **60**, 2077–2100.

Currier, R. M., Marsh, B. D. (2015) Mapping real time growth of experimental laccoliths: The effect of solidification on the mechanics of magmatic intrusion. *Journal of Volcanology and Geothermal Research*, **302**, 211–224.

Currier, R. M., Marsh, B. D., Mittal, T. (2010) Insights from analog gelatin experiments on the effect of bedding dip on sill morphology and crystal load. AGU Fall Meeting, #V43A-2343.

Dahm, T. (2000) On the shape and velocity of fluid-filled fractures on Earth. *Geophysical Journal International*, **142**, 181–192.

David, T. W. E., Priestley, R. E. (1914) Glaciology, physiography, and tectonic geology of South Victoria Land, with short notes on Paleontology by T. Griffith Taylor. British Antarctic Expedition 1907–1909, Reports on scientific investigations. *Geology*, **1**, 299.

Davidson, J. P., Charlier, B., Hora, J. M., Perlroth, R. (2005a) Mineral isochrons and isotopic fingerprinting: Pitfalls and promises. *Geology*, **33**(1), 29–32.

Davidson, J. P., Hora, J. M., Garrison, J. M., Dungan, M. A. (2005b) Crustal forensics in arc magmas. *Journal of Volcanology and Geothermal Research*, **140**, 157–170.

Davidson, J. P., Morgan, D. J., Charlier, B., Harlou, R., Hora, J. M. (2007) Microsampling and isotopic analysis of igneous rocks: Implications for the study of magmatic systems. *Annual Reviews Earth Planetary Sciences*, **35**, 273–311.

Denton, G. H., Sugden, D. E. (2005) Meltwater features that suggest Miocene ice-sheet overriding of the Transantarctic Mountains in Victoria Land, Antarctica. *Geografiska Annaler*, **87A**, 67–85.

Denton, G. H., Sugden, D. E., Marchant, D. R., Hall, B. L., Wilch, T. I. (1993) East Antarctic ice sheet sensitivity to pliocene climatic change from a Dry Valleys perspective. *Geografiska Annaler*, **75**(4), 155–204.

Drever, H. L., Johnston, R. (1967) Picritic minor intrusions. In Wyllie, P. J. (ed.) *Ultramafic and Related Rocks*. John Wiley and Sons, Inc., New York, 71–82.

Ducea, M. N. (2011) Fingerprinting orogenic delamination. *Geology*, **39**(2), 191–192.

Eales, H. V., Maier, W. D., Teigler, B. (1991) Corroded plagioclase feldspar inclusions in orthopyroxene and olivine of the lower and critical zones, western Bushveld complex. *Mineralogical Magazine*, **55**, 479–486.

Elburg, M., Goldberg, A. (2000) Age and geochemistry of Karoo dolerite dykes from northeast Botswana. *Journal of African Earth Sciences*, **31**, 539–554.

Elliot, D. H. (1975) Tectonics of Antarctica: A review. *American Journal of Science*, **275**, 45–106.

Elliot, D. H. (1996) The Hanson Formation: A new stratigraphical unit in Transantarctic Mountains, Antarctica. *Antarctic Science*, **8**, 389–394.

Elliot, D. H. (2013) The geological and tectonic evolution of the Transantarctic Mountains: a review. In Hambrey, M. J., Barker, P. F., Barrett, P. J., Bowman, V., Davies, B., Smellie, J. L., Tranter, M. (eds.) *Antarctic Palaeoenvironments and Earth-Surface Processes*. Geological Society, London, Special Publication no. **381**, 7–35.

Elliot, D. H., Fleming, T. H. (2000) Weddell triple junction: The principal focus of Ferrar and Karoo magmatism during initial breakup of Gondwana. *Geology*, **28**, 539–542.

Elliott, D. H., Fleming, T. H. (2004) Occurrence and dispersal of magmas in the Jurassic Ferrar Large Igneous Province, Antarctica. *Gondwana Research*, **7**, 223–237.

Elliot, D. H., Fleming, T. H. (2017) The Ferrar large igneous province: Field and geochemical constraints on supra-crustal (high-level) emplacement of the magmatic system. In Sensarma, S., Storey, B. C. (eds.) *Large Igneous Provinces from Gondwana and Adjacent Regions*. Geological Society, London, Special Publications, 463.

Elliot, D. H., Fleming, T. H. (2021) Ferrar large igneous province: Petrology, In: Smellie, J. L., Panter, K. S., Geyer, A. (eds.) *Volcanism in Antarctica: 200 Million Years of Subduction, Rifting and Continental Break-up*. Geological Society, London, Memoir 55, 93–120.

Elliot, D. H., Fleming, T. H., Kyle, P. R., Foland, K. A. (1999) Long distance transport of magmas in the Jurassic Ferrar large igneous province, Antarctica. *Earth and Planetary Science Letters*, **167**, 87–104.

Elliot, D. H., Hanson, R. E. (2001) Origin of widespread, exceptionally thick basaltic phreatomagmatic tuff breccia in the Middle Jurassic Prebble and Mawson Formations, Antarctica. *Journal of Volcanology and Geothermal Research*, **111**, 183–201.

Elliot, D. H., Larsen, D. (1993) Mesozoic volcanism in the Central Transantarctic Mountains, Antarctica: Depositional environment and tectonic setting. In: Findlay, R. H., Banks, M. R., Veevers, J. J., Unrug, R. (eds.) *Gondwana Eight: Assembly, Evolution and Dispersal, Tasmania, Australia*. A. A. Balkema, Rotterdam, 397–410.

Elliot, D. H., Larsen, D., Fannngs, C. M., Fleming, T. H., Vervoort, J. D. (2016) The lower Jurassic Hanson formation of the Transantarctic mountains: Implications for the Antarctic sector of the Gondwana plate margin. *Geological Magazine*, **154**, 777–803.

England, P. (1983) Constraints on extension of continental lithosphere. *Journal of Geophysical Research*, **88**, 1145–1152.

Ernst, R. E. (1990) *Magma Flow Directions in Two Mafic Proterozoic Dyke Swarms of the Canadian Shield, as Estimated Using Anisotropy of Magnetic Susceptibility Data. Mafic Dykes and Emplacement Mechanisms.* Balkema, Rotterdam, 231–235.

Ernst, R. E., Baragar, W. R. A. (1992) Evidence from magnetic fabric for the flow pattern of magma in the Mackenzie giant radiating dyke swarm. *Nature,* **356**, 511–513.

Ernst, R. E., Buchan, K. L. (1998) Giant radiating dike swarms: Their use in identifying pre-mesozoic large igneous provinces and mantle plumes. In: Mahoney, J. J., Coffin, M. F. (eds.) *Large Igneous Provinces: Continental, Oceanic, and Planetary Flood Volcanism,* Geophysical Monograph 100. American Geophysical Union, Washington, DC, 297–334.

Faure, G., Mensing, T. M. (2010) *The Transantarctic Mountains: Rocks, Ice, Meteorites and Water.* Springer, New York.

Ferrar, H. T. (1907) Report on the field geology of the region explored during the "Discovery" Antarctic Expedition 1901–1904. *National Antarctic Expedition, Natural History Reports,* **1**(1), 1–100.

Ferrar, H. T. (1925) The geological history of the Ross Dependency. *New Zealand Journal of Science and Technology,* **7**, 354–361.

Ferré, E. C., Bordarier, C., Marsh, J. S. (2002) Magma flow inferred from AMS fabrics in a layered mafic sill, Insizwa, South Africa. *Tectonophysics,* **354**, 1–23.

Ferré, E. C., Maes, S. M., Butak, K. C. (2009) The magnetic stratification of layered mafic intrusions: Natural examples and numerical models. *Lithos,* **111**, 83–94.

Ferris, J. K., Johnson, A., Storey, B. C. (1998) Form and extent of the Dufek intrusion, Antarctica, from newly compiled aeromagnetic data. *Earth and Planetary Science Letters,* **154**, 185–202.

Ferris, J. K., Storey, B. C., Vaughan, A. P. M., Kyle, P. R., Jones, P. C. (2003) The Dufek and Forrestal intrusions, Antarctica: A centre for Ferrar Large Igneous Province dike emplacement. *Geophysical Research Letters,* **30**, 1348.

Fiske, R. S., Jackson, E. D. (1972) Orientation and growth of Hawaiian volcanic rifts: The effect of regional structure and gravitational stresses. *Proceedings of the Royal Society of London A,* **329**, 299–326.

Fitzgerald, P. G. (1992) The Transantarctic Mountains of southern Victoria Land: The application of apatite fission track analysis to a rift shoulder uplift. *Tectonics,* **11**(3), 634–662.

Flanagan-Brown, R., Marsh, B. D. (2001) Fluid Dynamic Experiments on Mush Column Magmatism. American Geophysical Union Meeting, Spring, Montreal, abstract V23A-0686.

Fleming, T. H., Elliot, D. H., Jones, L. M., Bowman, J. R., Siders, M. A. (1992) Chemical and isotopic variations in an iron-rich lava flow from North Victoria Land, Antarctica: Implications for low-temperature alteration and the petrogenesis of Ferrar magmas. *Contributions to Mineralogy & Petrology,* **111**, 440–457.

Fleming, T. H., Foland, K. A., Elliot, D. H. (1995) Isotopic and chemical constraints on the crustal evolution and source signature of Ferrar magmas, North Victoria Land, Antarctica. *Contributions to Mineralogy & Petrology,* **121**, 217–236.

Fleming, T. H., Heimann, A., Foland, K. A., Elliot, D. H. (1997) 40Ar/39Ar geochronology of Ferrar Dolerite sills from the Transantarctic Mountains, Antarctica: Implications for the age and origin of the Ferrar magmatic province. *Geological Society of America Bulletin,* **109**, 533–546.

Foland, K. A., Marsh, B. D. (2005) Profiles of $^{87}Sr/^{86}Sr$ and $^{143}Nd/^{144}Nd$ as indicators of magma dynamics in the Ferrar dolerite magmatic system, McMurdo Dry Valleys, Antarctica (abstract). *Amererican Geophysics Union,* December.

Ford, A. B. (1976) Stratigraphy of the layered gabbroic Dufek intrusion, Antarctica: *U.S. Geological Survey Bulletin*, 1405-D, 36 p.

Ford, A. B., Himmelberg, G. R. (1991) Geology and crystallization of the Dufek intrusion. In: Tingey, R. J. (ed.) *The Geology of Antarctica*, Oxford Monographs on Geology and Geophysics, 17. Clarendon Press, Oxford. 175–214.

Ford, A. B., Kistler, R. W., White, L. D. (1986) Strontium and oxygen isotope study of the Dufek intrusion. *Antarctic Journal of the United States*, **21**(5), 63–66.

Ford, A. B., Nelson, S. W. (1972) Density of the stratiform Dufek intrusion, Pensacola Mountains, Antarctica. *Antarctic Journal of the United States*, **7**(5), 147–149.

Galland, O., Cobbold, P. R., Hallot, E., de Bremmond d'Ars, J., DeLavaud, G. (2006) Use of vegetable oil and silica powder for scale modeling of magmatic intrusion in a deforming brittle crust. *Earth and Planetary Science Letters*, **243**, 786–804.

Garcia, M. O., Hulsebosch, T. P., Rhodes, J. M. (1995) Olivine-rich submarine basalts from the southwest rift zone of Mauna Loa volcano: Implications for magmatic processes and geochemical evolution. In: Rhodes, J. M., Lockwood, J. P. (eds.) *Mauna Loa Revealed: Structure, Composition, History, and Hazards*. American Geophysical Union, Washington, DC, 219–239.

Garcia, M. O., Pietruszka, A. J., Rhodes, J. M. (2003) A petrologic perspective of Kilauea volcano's summit magma reservoir. *Journal of Petrology*, **44**, 2313–2339.

Gibb, F. G. F. (1968) Flow differentiation in the xenolithic ultrabasic dykes of the Cuillins and the Strathaird Peninsula, Isle of Skye, Scotland. *Journal of Petrology*, **9**, 411–443.

Gibb, F. G. F., Henderson, C. M. B. (1996) The Shiant Isles Main Sill: Structure and mineral fractionation trends. *Mineralogical Magazine*, **60**, 67–97.

Gottfried, D., Froelich, A. J. (1985) Geochemical and petrologic features of some mesozoic diabase sheets in the Northern Culpeper basin. *US Geological Survey Circular*, **946**, 86–90.

Gottfried, D., Froelich, A. J. (1988) Variations of palladium and platinum contents and ratios in selected Early Mesozoic tholeiitic rock associations in the Eastern United States. *US Geological Survey Bulletin*, **1776**, 332–340.

Gould, L. M. (1935) Structure of the Queen Maud Mountains, Antarctica. *Geological Society of America Bulletin*, **46**, 973–984.

Goulty, N. R. (2005) Emplacement mechanism of the Great Whin and Midland Valley dolerite sills. *Journal Geological Society of London*, **162**, 1047–1056.

Grapes, R. H., Reid, D. L. (1971) Rhythmic layering and multiple intrusion in the Ferrar Dolerite of South Victoria Land, Antarctica. *New Zealand Journal of Geology and Geophysics*, **14**(3), 600–604.

Grapes, R. H., Reid, D. L., McPherson, J. G. (1974) Shallow dolerite intrusion and phreatic eruption in the Allan Hills region, Antarctica. *New Zealand Journal of Geology and Geophysics*, **17**(3), 563–577.

Green, D. H., Ringwood, A. E. (1967) An experimental investigation of the gabbro to eclogite transformation and its petrological applications. *Geochimica et Cosmochimica Acta*, **31**, 767–833.

Grindley, G. W. (1963) The geology of the Queen Alexandra Range, Beardmore Glacier, Ross dependency, Antarctica; with notes on the correlation of Gondwana sequences. *New Zealand Journal of Geology and Geophysics*, **6**(3), 307–347,

Grindley, G. W., Warren, G. (1964) Stratigraphic nomenclature and correlation in the western Ross Sea region, Antarctica. In: Adie, R. J. (ed.) *Antarctic Geology. Proceedings of the First International [SCAR] Symposium of Antarctic Geology*, Capetown 16–21 September 1963. North Holland, Amsterdam, 314–333.

Gunn, B. M. (1962) Differentiation in Ferrar Dolerites, Antarctica. *New Zealand Journal of Geology and Geophysics*, **5**, 820–863.

Gunn, B. M. (1963) Layered intrusions in the Ferrar dolerites. *Mineralogical Society of America*, Special Paper, **1**, 124–133.

Gunn, B. M. (1966) Modal and element variation in Antarctic tholeiites. *Geochimica et Cosmochimica Acta*, **30**, 881–920.

Gunn, B. M., Warren, G. (1962) Geology of Victoria Land between the Mawson and Mulock Glaciers, Antarctica. *Bulletin of the Geological Survey of New Zealand*, **71**, 157.

Gunnarsson, B., Marsh, B. D., Taylor, H. P., Jr. (1998) Geology and petrology of postglacial silicic lavas from the SW part of the Torfajokull central volcano, Iceland. *Journal Volcanology and Geothermal Research*, **83**, 1–45.

Hamilton, W. B. (1964) Diabase sheets differentiated by liquid fractionation, TaylorGlacier region, south Victoria Land. In: Adie, R. J. (ed.) *Antarctic Geology*. North-Holland Publishing Company, Amsterdam, 442–454.

Hamilton, W. B. (1965) Diabase sheets of the Taylor Glacier region, Victoria Land, Antarctica. US Geological Survey, Professional Papers, 456-B, 71 p.

Hansen, D. M., Cartwright, J. A. (2006) Saucer-shaped sill with lobate morphology revealed by 3D seismic data: Implications for resolving a shallow-level sill emplacement mechanism, *Journal Geological Society*, **163**, 509–523.

Harris, C., le Roux, P., Cochrane, R., Martin, L., Duncan, A. R., Marsh, J. S., le Roex, A. P., Class, C. (2015) The oxygen isotope composition of Karoo and Etendeka picrites: High δ18O mantle or crustal contamination? *Contributions to Mineralogy and Petrology*, **170**, 8.

Harrowfield, D. L. (1995) *Vanda Station: History of an Antarctic Outpost 1963–1995*. New Zealand Antarctic Society Inc., New Zealand.

Hattori, K., Muehlenbachs, K. (1982) Oxygen isotope ratios of the Icelandic crust. *Journal of Geophysical Research*, **87**, 6559–6565.

Hayes. B., Lissenberg, C. J., Bedard, J. H., Beard, C. (2015) The geochemical effects of olivine slurry replenishment and dolostone assimilation in the plumbing system of the Franklin Large Igneous Province, Victoria Island, Arctic Canada. *Contributions to Mineralogy Petrology*, **169**, 22–48.

Hayman, P., Campbell, I. H., Cas, R. A. F., Squire, R. J., Doutch, D., Outhwaite, M. (2021) Differentiated Archean dolerites: Igneous and emplacement processes that enhance prospectivity for orogenic gold. *Economic Geology*, **116**(8), 1949–1980.

Heimann, A., Fleming, T. H., Elliot, D. H., Foland, K. A. (1994) A short interval of Jurassic continental flood basalt volcanism in Antarctica as demonstrated by 40Ar/39Ar geochronology. *Earth and Planetary Science Letters*, **121**, 19–41.

Heinonen, A., Kivisaari, H., Michallik, R. M. (2020) High-aluminum orthopyroxene megacrysts (HAOM) in the Ahvenisto complex, SE Finland, and the polybaric crystallization of massif-type anorthosites. *Contributions to Mineralogy and Petrology*, **175**, 1–25.

Heinonen, J. S., Luttinen, A. V., Spera, F. J., Vuori, S. K., Bohrson, W. A. (2021) Serial interaction of primitive magmas with felsic and mafic crust recorded by gabbroic dikes from the Antarctic extension of the Karoo large igneous province. *Contributions to Mineralogy and Petrology*, **176**, 28.

Hellweg, M. (2000) Physical models for the source of Lascar's harmonic tremor. *Journal of Volcanology and Geothermal Research*, 101, 183–198.

Henry, C. D. (2009) Uplift of the Sierra Nevada, California. *Geology, Geological Society of America*, 37, 575–576.

Hergt, J. M., Brauns, C. M. (2001) On the origin of Tasmanian dolerites. *Australian Journal of Earth Science*, **48**(4), 543–549.

Hergt, J. M., Chappell, B. W., Faure, G., Mensing, T. (1989a) The geochemistry of Jurassic dolerites from Portal Peak, Antarctica. *Contributions to Mineralogy and Petrology*, **102**, 298–305.

Hergt, J. M., Chappell, B. W., McCulloch, T. M., Mcdougall, I., Chivas, A. R. (1989b) Geochemical and isotopic constraints on the origin of the Jurassic dolerites of Tasmania. *Journal of Petrology*, **30**, 341–383.

Hergt, J. M., Peate, D. W., Hawkesworth, C. J. (1991) The petrogenesis of Mesozoic Gondwana low-Ti flood basalts. *Earth and Planetary Science Letters*, **105**, 134–148.

Hersum, T .G., Marsh, B. D., Simon, A. C. (2007) Contact partial melting of granitic country rock, melt segregation, and re-injection as dikes into Ferrar dolerite sills, McMurdo Dry Valleys, Antarctica. *Journal of Petrology*, **48**, 2125–2148.

Herzberg, C. T., Zhang, J. (1996) Melting experiments on anhydrous peridotite KLB-1: Composition of magmas in the upper mantle and transition zone. *Journal of Geophysical Research*, **101**, 8271–8295.

Hildreth, W., Moorbath, S. (1988) Crustal contributions to arc magmatism in the Andes of Central Chile. *Contributions to Mineralogy and Petrology*, **98**, 455–489.

Hill, D. P. (1977) A model for earthquake swarms. *Journal of Geophysical Research*, **82**, 1347–1352

Himmelberg, G. R., Ford, A. B. (1975) Petrologic studies of the Dufek intrusion-iron-titanium oxides. *Antarctic Journal of the United States*, **10**(5), 241–244.

Himmelberg, G. R., Ford, A. B. (1976) Pyroxenes of the Dufek intrusion, Antarctica. *Journal of Petrology*, **17**(2), 219–243.

Himmelberg, G. R., Ford, A. B. (1977) Iron-titanium oxides of the Dufek intrusion, Antarctica. *American Mineralogist*, **62**, 623–633.

Himmelberg, G. R., Ford, A. B. (1983) Composite inclusion of olivine gabbro and calc-silicate rock in the Dufek intrusion, a possible fragment of concealed contact zone. *Antarctic Journal of the United States*, **18**(5) 1–4.

Hinze, W. J., Chandler, V. W. (2020) Reviewing the configuration and extent of the Midcontinent rift system. *Precambrian Research*, **342**, 18.

Hirata, T. (1998) Fracturing due to fluid intrusion into viscoelastic materials. *Physical Review E*, **57**(2), 1772–1779.

Huber, N. K. (1981) Amount and timing of late cenozoic uplift and tilt of the central Sierra Nevada, California: Evidence from the Upper San Joaquin River Basin. *Geological Survey Professional Paper*, 1197, 28 p.

Hunter, R. H. (1987) Textural equilibrium in layered igneous rocks. In: Parsons, I. (ed.) *Origins of Igneous Layering*. D. Reidel Publishing Company, Dordrecht, 473–503.

Hyndman, D. W., Alt, D. (1987) Radial dikes, laccoliths, and gelatin models. *Journal of Geology*, 95, 763–774.

Irvine, T. N. (1980) Magmatic density currents and cumulus processes. *American Journal of Science*, **280-A**, 1–58.

Irvine, T. N. (1987) Layering and related structures in the Duke Island and Skaergaard intrusions: Similarities, differences, and origins. In: Parsons, I. (ed.) *Origins of Igneous Layering*. D. Reidel Publishing Company, Dordrecht, 185–245.

Ivanov, A. V., Meffre, S., Thompson, J., Corfu, F., Kamenetsky, V. S., Kamenetsky, M. B., Demonterova, E. I. (2017) Timing and genesis of the Karoo-Ferrar large igneous province: New high precision U-Pb data for Tasmania confirm short duration of the major magmatic pulse. *Chemical Geology*, **455**, 32–43.

Jaeger, H. M., Nagel, S. R. (1992) Physics of the granular state. *Science*, **255**, 1523–1531.

Jaeger, H. M., Nagel, S. R., Behringer, R. P. (1996) The physics of granular materials. *Physics Today*, **49**, 32–38.

Jaeger J. C., Joplin, G. (1955) Rock magnetism and the differentiation of dolerite sills. *Geological Society of Australia*, **2**, 1–19.

Jerram, D. A. (2005) Hot and cold in the Dry Valleys, Antarctica. *Geoscientist*, 15(9), 4–15.

Jerram, D. A., Davis, G. R., Mock, A., Charrier, A., Marsh, B. D. (2010) Quantifying 3D crystal populations, packing and layering in shallow intrusions: A case study from the Basement Sill, Dry Valleys, Antarctica. *Geosphere*, **6**(5), 537–548.

Jullien, R., Meakin, P., Pavlovitch, A. (1992) Three-dimensional model for a particle-size segregation by shaking. *Physical Review Lettsers*, **89**, 640–643.

Kavanagh, J. L., Menand, T., Sparks, R. S. J. (2006) An experimental investigation of sill formation and propagation in layered elastic media. *Earth and Planetary Science Letters*, **245**, 799–813.

Kibler, E. M. (1981) Peterogenesis of two Ferrar Dolerite sills, Roadend Nunatak, Transantartic Mountains, Antarctica. MS Thesis, The Ohio State University, Columbus, OH.

Kistler, R. W., White, L. D., Ford, A. B. (2000) Strontium and oxygen isotopic data and age for the layered gabbroic Dufek intrusion, Antarctica. *United States Geological Survey*. Open-File Report 00-133, 29 pp.

Klein, E. M., Langmuir, C. H. (1987) Global correlation of ocean ridge basalt chemistry with axial depth and crustal thickness. *Journal of Geophysical Research*, 92, 8089–8115.

Klemme, S., O'Neil, H. StC. (2000) The effect of Cr on the solubility of Al in Orthopyroxene: Experiments and thermodynamic modeling. *Contributions to Mineralogy and Petrology*, 140, 84–98.

Knight, J. B., Jaeger, H. M., Nagel, S. R. (1993) Vibration-induced size separation in granular media: The convection connection. *Physical Review Letters*, **70**, 3728–3731.

Kogiso, T. (2007) A geochemical and petrological view of mantle plume. In: Yuen, D., Maruyama, S., Karato, S., Windley, B. F. (eds.) *Superplumes: Beyond Plate Tectonics*. Springer, Dordrecht, 165–186.

Komar, P. D. (1976) Phenocryst interactions and the velocity profile of magma flowing through dikes or sills. *Geological Society of America*, **87**, 1336–1342.

Korenaga, J., Kelemen, P. B. (1998) Melt migration through the oceanic lower crust: A constraint from melt percolation modeling with finite solid diffusion. *Earth and Planetary Science Letters*, **156**, 1–11.

Korsch, R. J. (1984) The structure of Shapeless Mountain, Antarctica, and its relation to Jurassic igneous activity. *New Zealand Journal of Geology and Geophysics*, **27**, 487–504.

Koyaguchi, T., Takada, A. (1994) An experimental study on the formation of composite intrusions from zoned magma chambers. *Journal of Volcanology and Geothermal Research*, **59**, 261–267.

Kyle, P. R., Cole, J. W. (1974) Structural control of volcanism in the McMurdo volcanic group, Antarctica. *Bulletin Volcanologique*, **38**, 16–25.

Kyle, P. R., Pankhurst, R. J., Bowman, J. R. (1983) Isotopic and chemical variations in Kirkpatrick Basalt group rocks from southern Victoria Land. In: Oliver, R. L., James, P. R., Jago, J. B. (eds.) *Antarctic Earth Science*. Australian Academy of Sciences, Canberra, 234–237.

Lansing, A. (1959) *Endurance: Shackleton's Incredible Voyage*. Hodder and Stoughton, London.

Larson, P. B., Marsh, B. D. (2005) Oxygen isotopes of orthopyroxene and plagioclase from the dais layered intrusion of the Ferrar dolerite magmatic system; McMurdo Dry Valleys, Antarctica (abst.). *American Geophysics Union*, December.

Leat, P. T. (2008) On the long-distance transport of Ferrar magmas. *Geological Society, London, Special Publications*, **302**, 45–61.

Lewis, A. R., Marchant, D. R., Kowalewski, D .E., III, Baldwin, S. L., Webb, L. E. (2007) The age and origin of the Labyrinth, western Dry Valleys, Antarctica: Evidence for extensive middle Miocene subglacial floods and freshwater discharge to the Southern Ocean. *Geology*, **34**(7), 513–516.

Lister, J. R. (1991) Steady solutions for feeder dykes in a density-stratified lithosphere. *Earth and Planetary Science Letters*, **107**, 233–242.

Macdonald, K. C. (1986) The crest of the mid-atlantic ridge: Models for crustal generation processes and tectonics. In: Vogt, P. R., Tucholke, B. E. (eds.) *The Western North Atlantic Region. The Geology of North America*. Geological Society of America, 51–68.

Macdonald, K. C. (2019) *Mid-Ocean Ridge Tectonics, Volcanism, and Geomorphology*. *Encyclopedia of Ocean Sciences*, 3rd ed. Elsevier, Boulder, CO, 405–419.

MacLeod, C., Tyler, P. A., Tyler, P. A., Walker, C. (eds.) (1996) Tectonic, magmatic, hydrothermal and biological segmentation of mid-ocean ridges. *Geological Society Special Publication No. 118*. Geological Society of London, 288 pp.

Major, J. J., Pierson, T. C. (1992) Debris flow rheology: Experimental analysis of fine-grained slurries. *Water Resources Research*, **28**, 841–857.

Makse, H. A., Havlin, S., King, P. R., Stanley, H. E. (1997) Spontaneous stratification in granular mixtures. *Nature*, **286**, 379–382.

Malthe-Sørenssen, A., Planke, S., Svensen, H., Jamtveit, B. (2004) Formation of saucer-shaped sills. In: Breitkreuz, C., Petford, N. (eds.) *Physical Geology of High-Level Magmatic Systems*. Geological Society Special Publication, London, 234, 215–227.

Mangan, M. T, Marsh, B. D. (1992) Solidification front fractionation in phenocryst-free sheet-like magma bodies. *Journal of Geology*, **100**, 605–620.

Mangan, M. T., Marsh, B. D., Froelich, A. J., Gottfried, D. (1993) Emplacement and differentiation of the York Haven diabase sheet, Pennsylvania. *Journal of Petrology*, **34**, 1271–1302.

Marsh, B. D. (1976) Some Aleutian andesites: Their nature and source. *Journal of Geology*, 84, 27–45.

Marsh, B. D. (1979a) Island Arc Volcanism. *American Scientist*, **67**, 61–172.

Marsh, B. D. (1979b) Island Arc Development: Some observations, experiments and speculations. *Journal of Geology*, **87**, 687–713.

Marsh, B. D. (1981) On the crystallinity, probability of occurrence, and rheology of lava and magma. *Contributions to Mineralogy and Petrology*, **78**, 85–98.

Marsh, B. D. (1982) On the mechanics of igneous diapirism, stoping, and zone melting. *American Journal of Science*, **282**, 808–855.

Marsh, B. D. (1984) Mechanics and energetics of magma formation and ascension. In: Boyd, J. (ed.) *Explosive Volcanism, Inception, Evolution, and Hazards*. National Academies Press, Washington, DC, 67–83.

Marsh, B. D. (1989a) Magma chambers. *Annual Reviews of Earth and Planetary Sciences*, **17**, 439–474.

Marsh, B. D. (1989b) On convective style and vigor in sheet-like magma chambers. *Journal of Petrology*, **30**(3), 479–530.

Marsh, B. D. (1996) Solidification fronts and magmatic evolution. *Hallimond Lecture*. *Mineralogical Magazine*, **60**, 5–40.

Marsh, B. D. (2002) On bimodal differentiation by solidification front instability in basaltic magmas, I: Basic mechanics. *Geochimica et Cosmochimica Acta*, **66**, 2211–2229.

Marsh, B. D. (2004) A magmatic mush column Rosetta Stone: The McMurdo Dry Valleys of Antarctica. *EOS Transactions American Geophysical Union*, **85**(47), 497–502.

Marsh, B. D. (2006) Dynamics of magma chambers. *Elements*, **2**, 287–292.

Marsh, B. D. (2013) On some fundamentals of igneous petrology. *Contributions to Mineralogy and Petrology*, **166**, 665–690.

Marsh, B. D. (2015) *Magmatism, magma, and magma chambers*. In Watts, A. B. (ed.) *Treatise on Geophysics, Earth's Crust*, 2nd ed. Elsevier, Amsterdam, Chapter 6, 276–333.

Marsh, B. D., Coleman, N. M. (2009) Magma flow and interaction with waste packages in a geologic repository at Yucca Mountain, Nevada. *Journal of Volcanology and Geothermal Research*, **182**, 76–96.

Marsh, B. D., Gunnarsson, B., Congdon, R., Carmody, R. (1991) Hawaiian basalt and Icelandic rhyolite: Indicators of fundamental igneous processes. *Geologische Rundschau*, **80**, 481–510.

Mathez, E. A. (2005) Cold fire in Antarctica's Dry Valleys. *Natural History*, July/August, 26–31.

Mathez, E. A, McCallum, I. S., Marsh, B. D. (2005) On the mechanism of layering in the Dais Intrusion, McMurdo Dry Valleys, Antarctica. *American Geophysical Union Meeting*, San Francisco, abstract V23A-0686.

Mawson, D. (1916) Petrology of rock collections from the mainland of South Victoria Land. Report of British Antarctic Expedition. 1907–09. *Geology*, **2**(13), 201–234.

McDougal, I. (1962) Differentiation of the Tasmanian dolerites: Red Hill dolerite–granophyre association. *Geological Society of America Bulletin*, 73, 279–316.

McElroy, C. T., Rose, G. (1987) Geology of the Beacon Heights area, Southern Victoria Land, Antarctica. 1:50–000. *New Zealand Geological Survey miscellaneous series map 15 (1 sheet) and notes*. Wellington: Department of Scientific and Industrial Research.

McGinnis, L. D. (1981) Dry Valley drilling project. *Antarctic Research Serie*s. American Geophysical Union, 33, 465 p.

McKelvey, B. C., Webb, P. N. (1959) Geological investigations in South Victoria Land, Antarctica. Part 2; geology of Upper Taylor Glacier Region. *New Zealand Journal of Geology and Geophysics*, **2**(4), 718–728.

McKelvey, B. C., Webb, P. N. (1962) Geological investigations in southern Victoria Land, Antarctica. *New Zealand Journal of Geology and Geophysics*, **5**(1), 143–162.

McKelvey, B. C., Webb, P. N., Gorton, M. P., Kohn, B. P. (1970) Stratigraphy of the Beacon Supergroup between the Olympus and Boomerang ranges, south Victoria Land, Antarctica. *Nature*, **227**: 1126–1128.

McKelvey, B. C., Webb, P. N., Kohn, B. P. (1977) Stratigraphy of the Taylor and Lower Victoria Groups (Beacon Supergroup) between the Mackay Glacier and Boomerang Range, Antarctica. *New Zealand Journal of Geology and Geophysics*, **20**, 813–863.

McKenzie, D. (1984) The generation and compaction of partially molten rock. *Journal of Petrology*, **25**, 713–765.

Menand, T. (2008) The mechanics and dynamics of sills in layered elastic rocks and their implications for the growth of laccoliths and other igneous complexes. *Earth and Planetary Science Letters*, **267**, 93–99.

Mensing, T. M., Faure, G., Jones, L. M., Bowman, J. R., Hoefs, J. (1984) Petrogenesis of the Kirkpatrick Basalt, Solo Nunatak, northern Victoria Land, Antarctica, based on isotopic compositions of strontium, oxygen, and sulfur. *Contributions to Mineralogy and Petrology*, 87, 101–108.

Middleton, G. V. (1970) Experimental studies related to problems of flysch sedimentation. The Geological Association of Canada, Special Paper Number 7, 253–272.

Middleton, G. V. (1993) Sediment deposition from turbidity currents. In: Wetherill, G. W., Albee, A. L., Burke, K. C. (eds.) *Annual Review of Earth and Planetary Sciences*. Annual Reviews, Inc., Palo Alto, CA, 21, 89–114.

Mix, H. T., Ibarra, D. E., Mulch, A., Graham, S. A., Chamberlain, C. P. (2016) A hot and high Eocene Sierra Nevada. *Geological Society of America Bulletin*, **128**, 531–542.

Molzahn, M., Reisberg, L., Worner, G. (1996) Os, Sr, Nd, Pb, O isotope and trace element data from the Ferrar flood basalts, Antarctica: Evidence for an enriched subcontinental lithospheric source. *Earth and Planetary Science Letters*, **144**, 529–546.

Moritomi, H., Iwase, T., Chiba, T. (1982) A comprehensive interpretation of solid layer inversion in liquid fluidised beds. *Chemical Engineering Science*, **37**, 1751–1757.

Moritomi, H., Yamagishi, T., Chiba, T. (1986) Prediction of complete mixing of liquid-fluidized binary solid particles. *Chemical Engineering Science*, **41**, 297–305.

Morrison, A. D., Reay, A. (1995) Geochemistry of the Ferrar dolerite sills and dykes at Terra Cotta Mountain, south Victoria Land, Antarctica. *Antarctic Science*, **7**, 73–85.

Mudrey, M. G. Jr., Torii, T., Harris, H. (1975) Geology of DVDP 13 – Don Juan Pond. In: Mudrey, M. G. Jr., McGinnis, L. D. (eds.) *Dry Valley Drilling Project*, Bulletin No. **5**, Northern Illinois University Press, DeKalb, IL, 78–93.

Muirhead, J. D., Airoldi, G., Rowland, J. V., White, J. D. L. (2012) Interconnected sills and inclined sheet intrusions control shallow magma transport in the Ferrar large igneous province, Antarctica. *Bulletin Geological Society of America*, **124**, 162–180.

Muirhead, J. D., Airoldi, G., White, J. D. L., Rowland, J. V. (2014) Cracking the lid: Sill-fed dikes are the likely feeders of flood basalt eruptions. *Earth and Planetary Science Letters*, **406**, 187–197,

Murata, K. J., Richter, D. H. (1966) The settling of olivine in Kilauean magma as shown by the lavas of the 1959 eruption. *American Journal of Science*, **264**, 194–203.

Nasr-el-Din, H. A., Shook, C. A., Colwell, J. (1987) The lateral variation of solids concentration in horizontal slurry pipeline flow. *International Journal of Multiphase Flow*, **13**(5), 661–670.

Nedderman, R. (1992) *Statics and Kinematics of Granular Materials*. Cambridge University Press, Cambridge.

Neumann, E. R., Svensen, H., Galerne, C. Y., Plamke, S. (2011) Multistage evolution of dolerites in the Karoo Large Igneous Province, central South Africa. *Journal of Petrology*, **52**, 959–984.

Nicolas, A. (1995) *The Mid-Oceanic Ridges: Mountains Below Sea Level*. Springer-Verlag, Berlin.

Oskarsson, N., Sigvaldason, G. E., Steinthorsson, S. (1982) A dynamic model of rift zone petrogenesis and the regional petrology of Iceland. *Journal of Petrology*, **23**, 28–74.

Ottino, J. M., Khakhar, D. V. (2000) Mixing and segregation of granular materials. *Annual Review of Fluid Mechanics*, **32**, 55–91.

Pertermann, M., Hirschmann, M. M. (2003a) Partial melting experiments on a MORB-like pyroxenite between 2 and 3GPa: Constraints on the presence of pyroxenite in basalt source regions from solidus location and melting rate. *Journal of Geophysical Research*, **108**, 2125–2142.

Pertermann, M., Hirschmann, M. M. (2003b) Anhydrous partial melting experiments on MORB-like eclogite: Phase relations, phase compositions and mineral-melt partitioning of major elements at 2–3GPa. *Journal of Petrology*, **44**, 2173–2201.

Petford, N., Mirhadizadeh, S. (2017) Image-based modeling of lateral magma flow: The Basement Sill, Antarctica. *Royal Society Open Science*, **4**, 161083.

Philpotts, A. R. (1992) A model for emplacement of magma in the Mesozoic Hartford basin. *Geological Society of America, Special Paper*, **268**, 137–148.

Philpotts, A. R., Asher, P. M. (1993). Wallrock melting and reaction effects along the Higganum Diabase Dike in Connecticut: Contamination of a continental flood basalt feeder. *Journal of Petrology*, **34**, 1029–1058.

Philpotts, A. R., Carroll, M., Hill, J. M. (1996) Crystal-mush compaction and the origin of pegmatitic segregation sheets in a thick flood-basalt flow in the Mesozoic Hartford Basin, Connecticut. *Journal of Petrology*, **37**, 811–836.

Philpotts, A. R., Martello, A. (1986) Diabase feeder dikes for the Mesozoic basalts in southern New England. *American Journal of Science*, **284**, 105–126.

Philpotts, A. R., Philpotts, D. E. (2005) Crystal-mush compaction in the Cohassett flood-basalt flow, Hanford, Washington. *Journal of Volcanology and Geothermal Research*, **145**, 192–206.

Polteau, S., Ferre, E. C., Planke, S., Neumann, E. R., Chevallier, L. (2008a) How are saucer-shaped sills emplaced? Constraints from the Golden Valley Sill, South Africa. *Journal of Geophysical Research*, **113**, B12104.

Polteau, S., Mazzini, A. Galland, O., Planke, S., Malthe-Sorenssen, A. (2008b) Saucer-shaped intrusions: Occurrences, emplacement and implications. *Earth and Planetary Science Letters*, **261**, 195–204.

Pouliquen, O., Delour, J., Savage, S. B. (1997) Fingering in granular flows. *Nature*, **386**, 816–817.

Presnall, D. C., Hoover, J. D. (1987) High pressure phase equilibrium constraints on the origin of mid-ocean ridge basalts. In: Mysen, B. O. (ed.) *Magmatic Processes: Physicochemical Principles*. Geochemical Society Special Publication, Washington, DC, No **1**, 75–90.

Prior, G. T. (1899) Petrographical notes on the rock specimens collected in Antarctic regions during the voyage of H.M.S. Erebus and Terror under Sir James Clark Ross in 1839–1943. *Mineralogical Magazine*, **12**, 69–91.

Prior, G. T. (1902) Report on the rock specimens collected by the Southern Cross Antarctic Expedition. *Report on Southern Cross Antarctic Expedition*. British Museum London, 321–332.

Prior, G. T. (1907) Report on the rock specimens collected during the "Discovery" Antarctic Expedition, 1901–4. National Antarctic Expedition, 1901–4. *Natural History*, **1**, Geology (Field Geology: Petrography), 101–140.

Puffer, J. H., Philpotts, A. R. (1988) Eastern North American quartz tholeiites: Geochemistry and petrology. In: Manspeizer, W. (ed.) *Triassic–Jurassic Rifting*. Elsevier, Amsterdam, 579–605.

Putirka, K., Condit, C. D. (2003) A cross-section of a magma conduit system at the margin of the Colorado Plateau. *Geology*, **31**, 701–704.

Pyne, A. R. (1984) Geology of the Mt. Fleming area, South Victoria Land, Antarctica. *New Zealand Journal of Geology and Geophysics*, **27**, 505–512.

Ragland, P. C., Cummins, L. E., Arthur, J. D. (1992) Compositional patterns for early Mesozoic diabases from South Carolina to Central Virginia. *Geological Society of America, Special Paper*, **268**, 309–332.

Riffenburgh, B. (2004) Shackleton's forgotten expedition: The voyage of the Nimrod. Bloomsbury, London.

Riley, T. R., Leat, P. T., Curtis, M. L., Millar, I. L., Duncan, R. A., Fazel, A. (2006) Early–Middle Jurassic dolerite dykes from Western Dronning Maud Land (Antarctica):

Identifying mantle sources in the Karoo Large Igneous Province. *Journal of Petrology*, **46**, 1489-1524.

Ringwood, A. E. (1975) *Composition and Petrology of the Earth's Mantle*. McGraw-Hill, New York.

Rivalta, E., Dahm, T. (2006) Acceleration of buoyancy-driven fractures and magmatic dikes beneath the free surface. *Geophysics Journal International*, **166**, 1424–1439.

Roper, S. M., Lister, J. R. (2005) Buoyancy-driven crack propagation from an overpressured source. *Journal of Fluid Mechanics*, **536**, 79–98.

Ross, J. C. (1847) *A Voyage of Discovery and Research in the Southern and Antarctic Regions during the Years 1939–1843*, 2 vols. Spttiswoode and Shaw, London.

Ross, P. S., White, J. D. L., McClintock, M. (2008) Geological evolution of the Coombs–Allan Hills area, Ferrar large igneous province, Antarctica: Debris avalanches, mafic pyroclastic density currents, phreatocauldrons. *Journal of Volcanology and Geothermal Research*, **172**, 38–60.

Rubin, A. M. (1991) Dikes vs. Diapirs in viscoelastic rock. *Earth and Planetary Science Letters*, **119**, 641–659.

Rudnick, R. L. (1995) Making continental crust. *Nature*, **378**, 571–578.

Rudnick, R. L., Gao, S. (2005) Composition of the continental crust. In: Rudnick, R. L. (ed.) *The Crust*, Treatise on Geochemistry, vol.3. Elsevier, Amsterdam, 1–64.

Ryan, M. P. (1987) Elasticity and contractancy of Hawaiian olivine tholeiite and its role in the stability and structural evolution of subcaldera magma reservoirs and rift system. Volcanism in Hawaii. *US Geological Survey Professional Paper*, **1350**, 1395–1447.

Ryan, M. P. (1988) The mechanisms and 3-dimensional internal structure of active magmatic systems – Kilauea Volcano, Hawaii. *Journal of Geophysical Research, Solid Earth & Planets*, **93**, 4213–4248.

Ryan, M. P. (1993) Neutral buoyancy and the structure of midocean ridge magma reservoirs. *Journal of Geophysical Research, Solid Earth*, **98**, 22321–22338.

Savage, S. B., Lun, C. K. K. (1988) Particle size segregation in inclined chute flow of dry cohesionless granular solids. *Journal of Fluid Mechanics*, **189**, 311–335.

Schouten, H., Klitgord, K. D., Whitehead, J. A. (1985) Segmentation of mid-ocean ridges. *Nature*, **387**, 225–229.

Scott, R. F. (1905) *The Voyage of the 'Discovery'*, 2 vols. Smith, Elder & Co., London.

Self, S., Keszthely, L. P., Thordarson, T. (2000) Discussion of: 'Pulsed inflation of pahoehoe lava flows for flood basalt emplacement', by Anderson, S. W., Stofan, E. R., Smrekar, S. E., Guest, J. E., Wood, B. [Earth Planet. Sci. Lett. 168 (1999) 7–18]. *Earth and Planetary Science Letters*, 179, 421–423.

Self, S., Thordarson, T, Keszthely, L. P., et al. (1996) A new model for the emplacement of Columbia River basalts as large inflated pahoehoe lava flow fields. *Geophysical Research Letters*, **25**, 2689–2692.

Seward, A. C. (1914) Antarctic fossil plants. In: British Antarctic ("Terra Nova") expedition, 1910. *Natural History Reports Geology* **1**, 1. London: British Museum (Natural History): 1–49

Shackleton, E. H. (1907) *The Heart of the Antarctic: The Farthest South Expedition 1907–1909*, 2 vols. William Heinemann, London.

Simkin, T. (1967) Flow differentiation in the picritic sills of North Skye. In: Wyllie, P. J. (ed.) *Ultramafic and Related Rocks*. John Wiley and Sons, New York, 64–69.

Simkin, T. (1993) Terrestrial volcanism in space and time. *Annual Reviews of Earth and Planetary Science*, **21**, 427–452.

Sinton, J. M., Detrick, R. S. (1992) Mid-ocean ridge magma chambers. *Journal of Geophysical Research*, **97**, 197–216.

Skelton, A. D., Graham, C. M., Bickle, M. J. (1995) Lithological and structural controls on regional 3-D fluid flow patterns during greenschist facies metamorphism of the Dalradian of the SW Scottish Highlands. *Journal of Petrology*, **36**(2), 563–586.

Smith, W. C. (1924) The plutonic and hypabyssal rocks of South Victoria Land. Natural History Report, British Antarctic Terra Nova Expedition 1910–1913. *Geology*, **1**, 167–227.

Spandler, C., Yaxley, G., Green, D. H., Rosenthal, A. (2007) Phase relations and melting of anhydrous K-bearing eclogite from 1200 to 1600°C and 3 to 5 GPa. *Journal of Petrology*, **49**(4), 771–795.

Stewart, D., Jr. (1951) On the mineralogy of Antarctica. *American Mineralogist*, **36**, 362–367.

Stolper, E. (1980) A phase diagram for mid-ocean ridge basalts: Preliminary results and implications for petrogenesis: *Contributions to Mineralogy and Petrology*, **74**, 13–27.

Storey, B. C. (1995) The role of mantle plumes in continental breakup: Case histories from Gondwanaland. *Nature*, **377**, 301–308.

Stroujkova, A., Malin, P. (2001) Multiple ruptures for Long Valley microearthquakes: A link to volcanic tremor? *Journal of Volcanology and Geothermal Energy*, **106**, 123–143.

Stump, E. (1995) *The Ross Orogen of the Transantarctic Mountains*. Cambridge University Press, New York.

Sugden, D., Denton, G. (2004) Cenozoic landscape evolution of the Convoy Range to MacKay Glacier area, Transantarctic Mountains: Onshore to offshore synthesis. *Geological Society of America Bulletin*, **116**, 840–857.

Sushchevskaya, N. M., Belyatsky,B. V., Leichenkov, G. L., Laiba, A. A. (2009) Evolution of the Karoo–Maud Mantle Plume in Antarctica and its influence on the magmatism of the early stages of Indian Ocean opening. *Geochemistry International*, **47**, 1–17.

Takada, A. (1994) Development of a subvolcanic structure by the interaction of liquid-filled cracks. *Journal of Volcanology and Geothermal Reaearch*, **62**, 207–224.

Taylor, G. (1916) *With Scott: The Silver Lining*. Smith, Elder & Co., London. (reprinted, 1997, Bluntisham – Erskine Press, Norfolk.)

Taylor, G. (1922) *The Physiography of the McMurdo Sound and Granite Harbor Region*. Harrison & Sons, London.

Taylor, H. P., Jr. (1977) Stable isotope studies of spreading centers and their bearing on the origin of granophyres and plagiogranites. *Colloques Internationaux du Centre National de la Recherchie Scientifique*, 149–165.

Taylor, H. P., Jr. (1987) Comparison of hydrothermal systems in layered gabbros and granites and the origin of low 18O magmas. In: Mysen, B.O. (ed.) *Magmatic Processes: Physiochemical Principles*. The Geochemical Society Special Publication, Washington, DC, No. 1, 337–357.

Teagle, D. A. H., Ildefonse, B., Blum, P, IODP Expedition 335 Scientists (2012) *IODP Expedition 335: Deep Sampling in ODP Hole 1256D*. Scientific Drilling, no. 13, April, 28–34.

Thompson, G. A. (1998) Deep mantle plumes and geoscience vision. *GSA Today*, April, 17–25.

Thomson, K., Hutton, D. H. W. (2004) Geometry and growth of sill complexes: Insights using 3D seismic from the North Rockall trough. *Bulletin of Volcanology*, **66**, 364–375.

Turnbull, I. M., Allibone, A. H., Forsyth, P. J., Heron, D. W. (1994) *Geology of the Bull Pass - St. Johns Range Area, Southern Victoria Land, Antarctica*. Institute of

Geological & Nuclear Sciences Geological Map 14. Institute of Geological & Nuclear Sciences Limited, Lower Hutt, New Zealand.

Tyler-Lewis, K. (2006) *The Lost Men: The Harrowing Saga of Shackleton's Ross Sea Party*. Penguin Books, New York.

Vallance, J. W. (1994) Experimental and field studies related to the behavior of granular mass flows and the characteristics of their deposits. Unpublished PhD Dissertation, Michigan Technological University, 197 p.

Walker, B. A., Jr., Bergantz, G. W., Otamendi, J. E., Ducea, M. N., Cristofolini, E. A. (2015) A MASH zone revealed: The mafic complex of the Sierra Valle Fértil. *Journal of Petrology*, **56**(9), 1863–1896.

Walker, D., Shibata, T., Delong, S. E. (1979) Abyssal tholeiites from the Oceanographer Fracture Zone II: Phase equilibria and mixing. *Contributions to Mineralogy and Petrology*, **70**, 111–125.

Watanbe, T., Masuyama, T., Nagaoka, K., Tahara, T. (2002) Analog experiments on magma-filled cracks: Competition between external stresses and internal pressure. *Earth Planets and Space*, **54**, 1247–1261.

Webb, P. N., McKelvey, B. C. (1959) Geological investigations in South Victoria Land, Antarctica. Part I – Geology of Victoria Valley. *New Zealand Journal of Geology and Geophysics*, **2**(1), 120–136.

White, J. D. L., Bryan, S. E., Ross, P.-S., Self, S., Thordarson, T. (2009) Physical volcanology of continental large igneous provinces: Update and review. In: Thordarson, T., Self, S., Larsen, G., Rowland, S. K., Hoskuldsson, A. (eds.) *Studies in Volcanology: The Legacy of George Walker*. Special Publications of IAVCEI. Geological Society, London, **2**, 291–321.

White, J. D. L., McClintock, M. K. (2001) Immense vent complex marks flood-basalt eruption in a wet failed rift: Coombs Hills, Antarctica. *Geology*, **29**, 935–938.

White, R. A., McCausland, W. A. (2019) A process-based model of pre-eruption seismicity patterns and its use for eruption forecasting at dormant stratovolcanoes. *Journal of Volcanology and Geothermal Research*, **382**, 267–297.

Wilch, T. I., Denton, G. H., Lux, D. R., McKintosh, W. C. (1993) Limited pliocene glacier extent and surface uplift in middle Taylor Valley, Anarctica. *Geografiska Annaler*, **75A**, 331–351.

Wilson, A. H. (1996) The great Dyke of Zimbabwe. In: Cawthorn, R. G. (ed.) *Layered Intrusions*. Elsevier, Amsterdam, 365–402.

Wilson, E. (1966) *Diary of the Discovery Expedition to the Antarctic regions 1901–1904*. Blandford Press, London.

Wójcik, M., Tejchman, J. (2009) Modeling of shear localization during confined granular flow in silos within non-local hypoplasticity. *Powder Technology*, **192**, 298–310.

Wolfe, E. W., Garcia, M. O., Jackson, D. B., Koyanagi, R. Y., Neal, C. A., Okamura, A. T. (1987) The Pu,u O,o eruption of Kilauea volcano, episodes 1–20, January 3, 1983 to June 8, 1984. *U.S. Geological Survey Professional Paper*, **1350**, 471–508.

Worst, B. G. (1960) The Great Dyke of Southern Rhodesia. *Southern Rhodesia Geological Survey Bulletin*, **47**, 234.

Wright, A. C., Kyle, P. R. (1990) McMurdo volcanic group, Western Ross Embayment. In: LeMasurier, W. E., Thomson, J. W. (eds.) *Volcanoes of the Antarctic Plate and Southern Oceans*, Antarctic Research Series. American Geophysical Union, Washington, DC, **48**, 97-134.

Wright, T. L. (1971) Chemistry of Kilauea and Mauna Loa lava in space and time. *Geological Survey Professional Paper 735*, 40.

Wright, T. L., Fiske, R. (1971) Origin of the differentiated and hybrid lavas of Kilauea volcano, Hawaii. *Journal of Petrology*, **12**, 1–65.

Wright, T. L., Marsh, B. D. (2016) Quantification of the intrusion process at Kilauea Volcano, Hawai'i. *Journal of Volcanology and Geothermal Research*, **328**, 34–44.

Zavala, K., Leitch, A. M., Fisher, G. W. (2011) Silicic segregations of the Ferrar dolerite sill, Antarctica. *Journal of Petrology*, **52**(10), 1927–1964.

Zieg, M. J., Marsh, B. D. (2005) The Sudbury igneous complex: Viscous emulsion differentiation of a superheated impact melt sheet. *Geological Society America Bulletin*, **117**, 1427–1450.

Zieg, M. J., Marsh, B. D. (2012) Multiple reinjections and crystal-mush compaction in the Beacon Sill, McMurdo Dry Valleys, Antarctica. *Journal of Petrology*, **53**(12), 2567–2591.

Zieg, M. J., Wallrich, B. M. (2018) Emplacement and differentiation of the Black Sturgeon Sill, Nipigon, Ontario: A principal component analysis. *Journal of Petrology*, **59**(19), 2385–2412.

# Index

Printed in the United States
by Baker & Taylor Publisher Services